ANALYTIC GEOMETRY

ANALYTIC GEOMETRY

Graphic Solutions Using MatLab Language

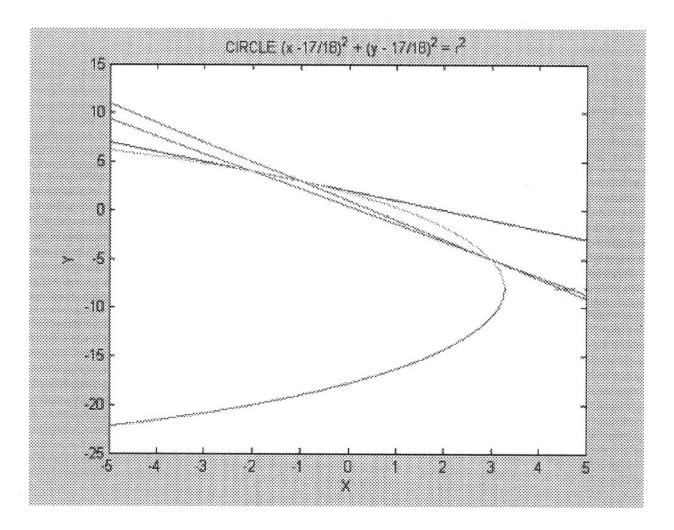

ING. MARIO CASTILLO

ISBN: Softcover 978-1-4633-7256-9
Ebook 978-1-4633-7257-6

This book was printed in the United States of America.

Rev. date: 05/11/2013

To order additional copies of this book, contact:
Palibrio LLC
1663 Liberty Drive
Suite 200
Bloomington, IN 47403
Toll Free from the U.S.A 877.407.5847
Toll Free from Mexico 01.800.288.2243
Toll Free from Spain 900.866.949
From other International locations +1.812.671.9757
Fax: 01.812.355.1576
orders@palibrio.com
504853

PREFACE

FOR THE SOLUTON OF THE PROBLEMS THIS BOOK INCLUDE ARE: THE COMMONLY SOLUTION USED IN THE ANALYTIC GEOMETRY SUBJET, AND THE GRAPHIC SOLUTIONS USING MATLAB LANGUAGE WITH THE PURPOSE HELP AT THE STUDENT VISUALIZE AND LEARN COMPUTER PROGRAMMING.

CONTENTS

SEGMENTS

PROBLEM 1. DRAW THE POINTS A(4,4), B(-4,4), C(-4,-4), D(4, -4) AND E(2^.5, 3^.5).

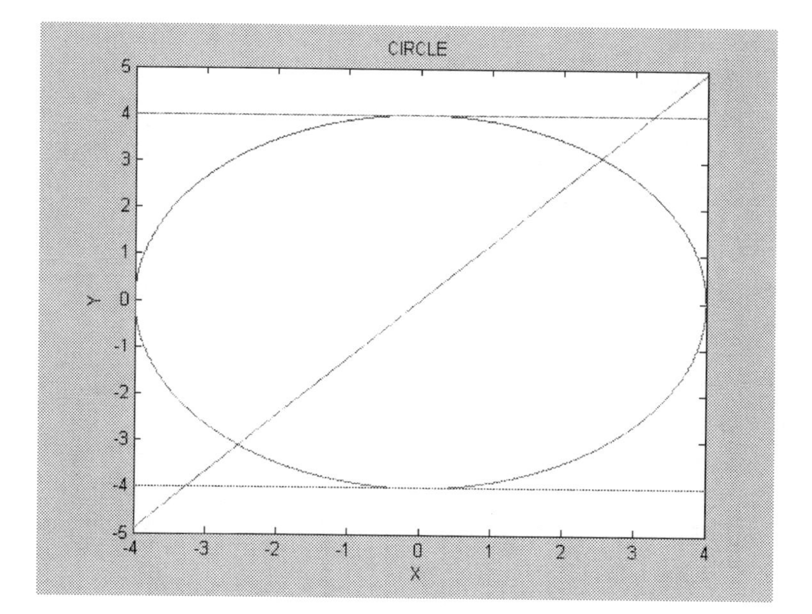


```
x = -4: .001: 4;
if (16 - x.*x) >= 0
y = sqrt(16 - x.*x);
end
if (16 - x.*x) >= 0
y1 = - sqrt(16 - x.*x);
end
```

```
x = -4: .001: 4;
y2 = ((3/2)^.5)*x;
x = -4: .001: 4;
y3 = -4;
x = -4: .001: 4;
y4 = 4;
plot(x,y, x, y1, x, y2, x,y3, x, y4);
xlabel('X');
ylabel('Y');
title('CIRCLE ');
```

PROBLEM 2. WHICH ARE THE ALGEBRAIC SIGNS OF THE COORDINATES IN EACH OF THE FOUR QUADRANTS?

QUADRANT I (+, +)

QUADRANT II (-, +)

QUADRANT III (-,-)

QUADRANT IV (+,-)

PROBLEM 3. IF ONE POINT IS OVER X AXLE, WHAT IS ITS ORDINATE? IF THE POINT IS OVER Y AXLE, WHAT IS IT'S ABSCISE? WHERE ARE SITUATED THE POINTS WHOSE ABSICE IS X = 1? AND WHERE THE POINTS HAVE Y = -2?.

a) y = 0;

b) x = 0;

c) OVER Y AXLE.

d) OVER X AXLE.

PROBLEM 4. DRAW THE POINTS A (0, -2), B (1, 1), C (2, 4) AND D (-1,-5) AND PROBE THAT THE POINTS ARE OVER A STRAIGTH LINE.

mAB = (1 + 2)/(1 – 0) = 3,   mCD = (-5 – 4)/(-1- 2) = (-9/-3) = 3

```
x = -1: .001: 2;
%(y- y1) = m(x - x1)
y = (3)*x +2;
plot(x,y);
xlabel('X');
ylabel('Y');
title('STRAIGTH LINE ');
```

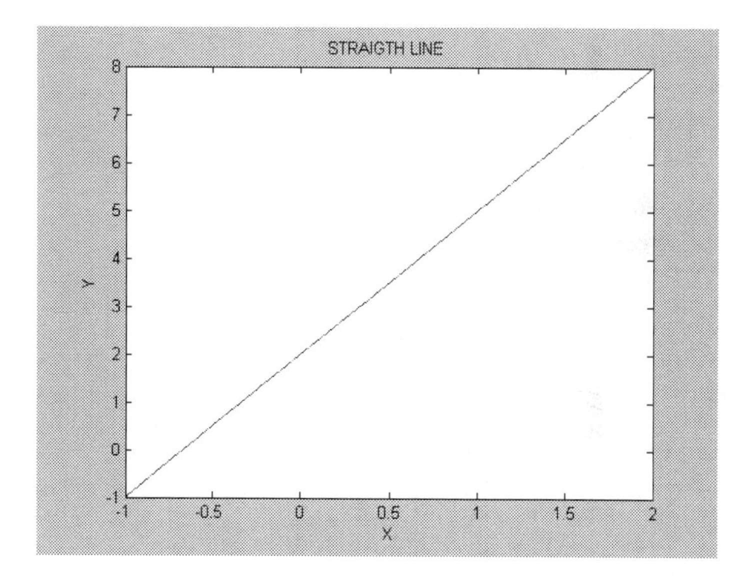

PROBLEM 5. PROBE GRAPHYCALLY THAT THE NEXT POINTS ARE SITUATED OVER ONE CIRCLE AND FIND ITS CENTER AND RADIUS.

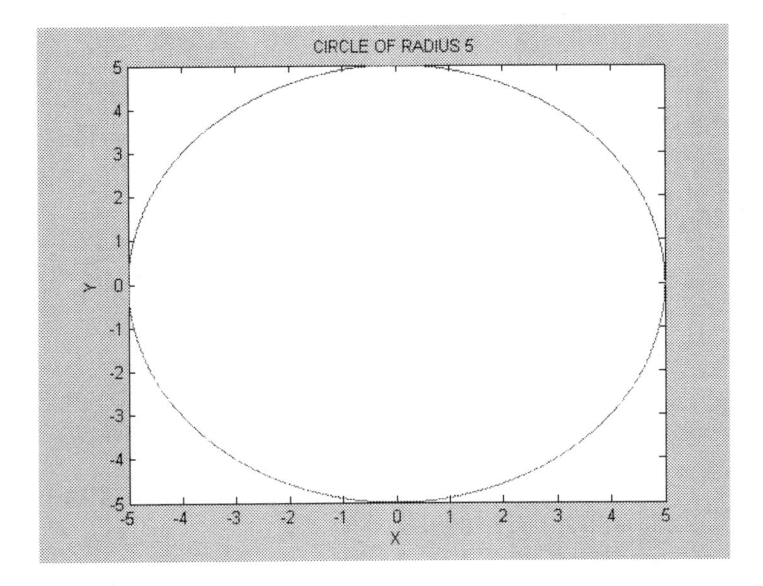


```
x = -5: .001: 5;
if (25 - x.*x) >= 0
y = sqrt(25 - x.*x);
end
if (25 - x.*x) >= 0
y1 = -sqrt(25 - x.*x);
end
plot(x,y, x, y1);
xlabel('X');
ylabel('Y');
title('CIRCLE OF RADIUS 5');
```

PROBLEM 6. DRAW THE POINTS A (1, 2) AND B (3, 4). SUPPOSE THAT THE HORIZONTAL BY POINT A INTERSECT THE VERTICAL BY POINT B

IN P. WHAT ARE THE COORDINATES OF POINT P? WHISH ARE THE LONGITUDE OF AP AND BP? CALCULATE THE DISTANCE AB

a) P (1,4) LONGITUDE AP = X2 –x1 = 0; y2 –y1 = 2; AP = 2 BP = x2-x1 = 2, y2 –y1 = 0;

BP = 2 DISTANCE AB = ((x2-x1)^2 + (y2-y1)^2)^.5 = (4 + 4)^.5 = 2(2)^.5

PROBLEM 7. FIND THE DISTANCE BETWEEN A (-2,-2) AND B (3, -4).

AB = ((-2 -3)^2+ (-2 +4)^2)^.5 = (25 + 4)^.5 = (29)^.5

DISTANCE BETHWEEN POINTS

PROBLEM 1. FIND PERIMETER OF THE TRIANGLE FORMED BY THE POINTS A (2, 3),

B(-3,3) AND C(1,1). AB = ((2 + 3)^2 + (3 – 3)^2)^.5 = 5

BC = ((1 + 3)^2+ (1 – 3)^2)^.5 = (16 + 4)^.5 = 2(5)^.5

CA = ((2 – 1)^2+ (3 – 1)^2)^.5 = (1 + 4)^.5 = (5)^.5

PERIMETER = AB + BC + CA = 5 + 3(5)^.5

PROBLEM 2. FIND THE COORDINATES OF THE POINTS WHOSE DISTANCES AT THE POINT A (2, 3) IS 4 AND ORDINATES ARE EQUAL AT 5.

Distance of A(2,3) is = 4 (x – 2)^2 + (y -3)^2 = 16, and Ordinate y = 5

(x – 2)^2 + (y -3)^2 = 16,

(x – 2)^2 + (y -3)^2 = 16

(x – 2)^2 + (5-3)^2 = 16

x^2 – 4x + 4 + 4 = 16

x^2 – 4x – 8 = 0

x = (-b +- (b^2 – 4ac))/2a = (4 +- ((16 + 4(1)(8))^.5)/2(1) =

x = (4 +- (48)^^.5)/2 = 2 +- 2(3)^.5

121 DEMOSTRATE THAT THE POINTS A(1, -2), B(4, 2) AND C(-3, -5) ARE THE VERTEX OF ONE ISOSEL TRIANGLE.

find the longitude of the sides AB and CA

AB = ((4-1)^2+ (-2 -2)^2)^.5 = (9 + 16)^.5 = 5

CA = ((-3 -1)^2 + (-5 +2)^2)^.5 = (16 + 9)^.5 = 5

122 DEMOSTRATE THAT THE POINTS A (0, 0), B (3, 1), C (1, -1), AND D (2, 2) ARE THE SIDES OF ONE PARALLELOGRAM.

BC = ((3 -1)^^2 + (1 +1)^2)^.5 = (4 + 4)^^.5 = 2(2)^.5

AD = (2^2 + 2^2)^^.5 =2(2)^.5

AB = ((3^2 + 1^2)) ^.5 = (10)^.5

CD = ((2- 1)^2 + (2 + 1)^2)^.5 = (10)^.5 THE FIGURE IS A PARALELOGRAM.

PROBLEM 129. KNOWN A (0, 0), B (1, 1), C (-1, 1) AND D (1,-2) FIND THE POINT IN WHICH THE PERPENDICULAR AT THE MIDLE SEGMENT AB, INTERSECT WITH THE PERPENDICULAR IN THE MIDLE POINT OF THE SEGMENT CD.

P1 IN AB P1(1/2, 1/2) P2 IN CD x1 = (-1 +1)/2 =0, y2(1-2)/2 = -1/2 P2(0, -1/2)

PROBLEM 130. FIND THA FEET OF THE PERPENDICULAR FROM THE POINT A (1, 2) AT THE STRAIGTH LINE JOINED BY THE POINTS B (2, 1) AND c (-1,-5).

SLOPE mBC =(y2 –y1)/(x2 – x1) = 6/3 =2 PERPENDICULAR SLOPE = -1/2

STRAIGTH LINE y –y1 = m(x – x1)

LINE BC (y - 1) = (2)(x -2), y - 1 =2 x -4, y =2x -3

PERPENDICULAR LINE (y – 2) = -1/2(x – 1), 2y -4 = -x + 1, y = -x /2+5/2

INTERSECTION SOLUTION OF THE TWO EQUATIONS y = 2x - 3 = -x/2 +5/2

2x -3 = - x/2 + 5/2, 5x/2 = 11/5, x = 22/5 - 3, y = 22/5 -15/5 = 7/5.

INTERSECTION POINT P(11/5, 7/5).

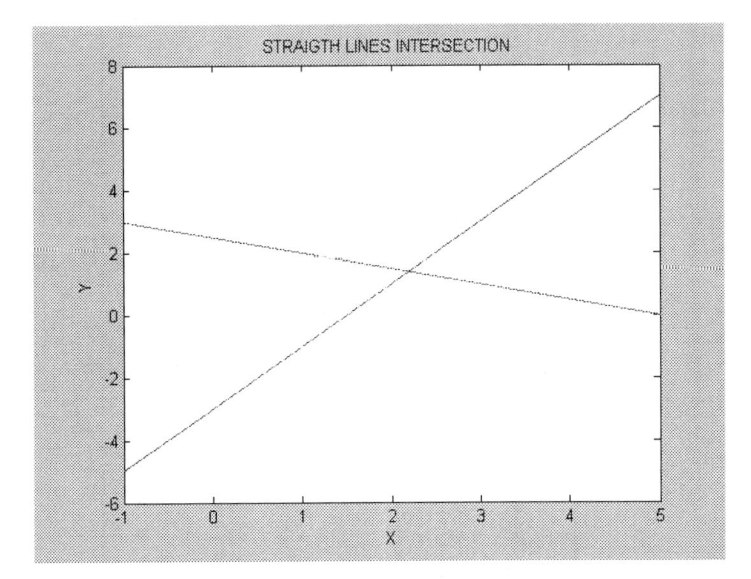


```
x = -1: .001: 5;
%(y- y1) = m(x - x1)
y = 2*x -3;
x1 = 5: -.001: -1;
y1 = - x1/2 + 5/2;
plot(x, y, x1, y1);
xlabel('X');
ylabel('Y');
title('STRAIGTH LINES INTERSECTION ');
```

VECTORS

PROBLEM 137. ONE VECTOR EQUAL [-3, 4] HAS BY ORIGIN THE POINT (1, -2) WHICH ARE THE COORDINATES OF IT'S EXTREME?

AB = [-3 4] = [x - 1, y + 2]

-3 = x – 1   y + 2 = 4

x = -2   y = 2   POINT (-2, 2)

PROBLEM 138. THE POINTS A (1, 2), B (-2, -1), C (3, -2) ARE THE VERTEX OF ONE TRIANGLE. FIND THE VECTORS JOIN THE VERTEX AND THE MIDDLE POINTS OF THE OPPOSITE SIDES.

BM1 = M1C ,

[x1 +2, y1 +1] = [3 –x1 , -2 – y1]

[x1 +2, y1 +1] = [3 –x1 , -2 – y1]

X1 + 2 = 3 – x1 y1 +1 = -2 – y1

2x1 = 1   2y1 = -3

x1 = 1/2   y1 = -3/2

P1(1/2, -3/2)

AP1 = [-1/2 -7/2]

AM2 = M2B,

[x2 - 1, y2 - 2] = [-2 - x2 , -1 – y2]

[x2 -1, y2 – 2] = [-2 – x2, -1 – y2]

2x2 = -1   2y2 = 1

x2 = -1/2   y2 = 1/2

P2 (-1/2, 1/2)

CP2 = [-7/2 5/2]

AM3 = M3C

[x3 - 1, y3 - 2] = [3 –x3 , -2 – y3]

2x3 = 4   2y3 = -0

X3 = 2   y3 = 0 P3 (2, 0)   BP3 = [4 1]

139. KNOWING A(2,3), B(-4,5) AND C(-2,3) FIND D SUCH THAT AB = CD.

(-4 -2)^2 + (5 – 3)^2 = (x +2)^2 + (y -3)^2 WE HAVE TWO EQUATIONS

36 + 4 = (x +2)^2 + (y – 3)^2 = 40

(x +2)^2 + (y – 3)^2 = 40 C(-2, 3) R = (40)^.5

AB = CD [-4 -2, 5 - 3] = [x +2, y -3]

-6 = x +2 2 = y – 3

X = -8 y = 5 D(-8, 5)

PROBLEM 140. THE MIDDLE POINT OF ONE SEGMENT IS C (1, 2), AND ONE OF IT'S EXTREMS HAS BY COORDINATES A (-3, 5) FIND THE COORDINATES OF THE ANOTHER EXTREM. AC = CD [1 + 3, 2 -5] = [x - 1, y - 2]

4 = -1 + x -3 = - 2 + y

X = 5 y =- 1 D(5, -1)

PROBLEM 141. IF THE AREA OF ONE TRIANGLE ABC IS CERO. DEMOSTRATE THE POINTS A(0,-2), B(1,1) AND C(3, 7) ARE SITUATED OVER ONE STRAIGTH LINE.

mAB = (1+2)/(1 -0) = 3, mBC = (7 – 1)/(3 – 1) = 3 THE SLOPES ARE EQUALS IS ONE STRAIGTH LINE (y –y1) = m (x – x1)

(y +2) = 3(x)

y = 3x -2

```
%(y- y1) = m(x - x1)
x = -5: .001: 5;
y = 3*x - 2;
plot(x, y);
xlabel('X');
ylabel('Y');
title('STRAIGTH LINE ');
```

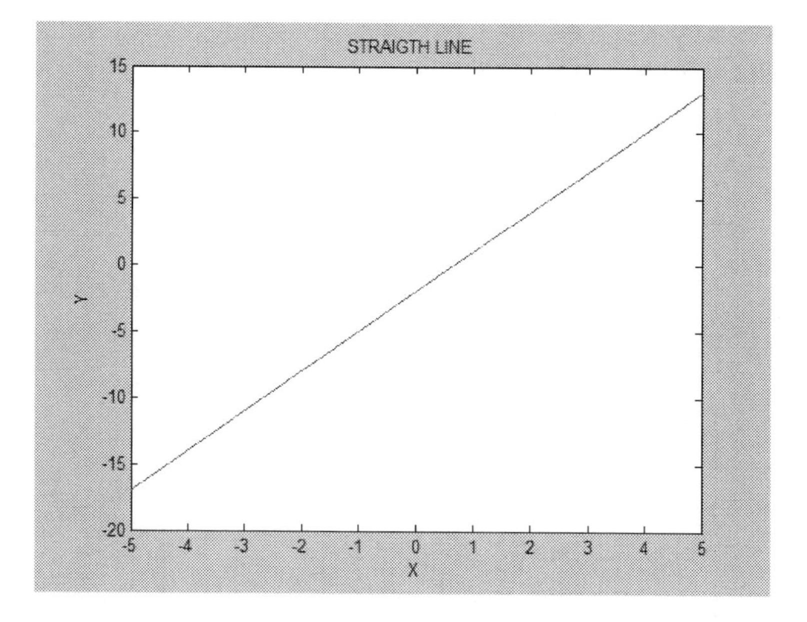

PROBLEM 142. FIND THE AREA OF THE QUADRILATER WITH VERTEX A (-2,-3), B (1, -2), C(3,4) AND D(-1,5).

AB = ((-2 -1)^2 + (-3 +2)^2)^.5 = (9 +1)^.5 = (10)^.5 Area1 = b*h/2 = (3)(1)/2 = 3/2

BC = ((3 – 1)^2 + (4 + 2)^2)^.5 = (4 + 36) = 2(10)^.5 Area2 = (2)(6)/2 = 6

CD = ((3 + 1)^2 + (4 -5)^2)^.5 = (16 + 1)^^.5 = (17)^.5 Area3 = (4)(1)/2 = 2

DA = ((-2 + 1)^^2 + (-3 -5)^^2)^.5= (1 + 64) = (65)^.5 Area4 = 8(1)/2 = 4

Area Total = 24 (3/2)

ADD AND SUBSTRACT OF VECTORS

PROBLEM 144. GIVEN THE POINTS: P(1, -3), Q(7, 1), R(-1,1) AND S(2,3) DEMOSTRATE PQ HAS THE SAME DIRECTION OF RS AND IT'S DOUBLE.

PQ = [7 – 1, 1 + 3] = [6 4] = 2[3 2]

RS = [2 +1, 3-1] = [3 2]

PROBLEM 145. GIVEN THE POINTS P1(2, -3), P2(-1,2) LOOK FOR OVER P1P2 THE POINT WITH DOUBLE DISTANCE OF P1 THAT P2.

P1P = (2/3)P1P2 , [x – 2 , y + 3] = (2/3)[-1 -2, 2+ 3]

X – 2 = 2 y + 3 = 10/3

X = 0 y =10/3 – 9/3 = 1/3 p(0, 1/3)

PROBLEM 146. FIND THE POINTS P AND Q OVER THE STRAIGTH LINE PASS BY P1 (2, -1) AND P2(-4, 5) SUCH THAT P1P = -(2/3)PP2 AND P1Q = -(2/3)P1P2.

P1P = -(2/3) PP2 P1Q = - (2/3)P1P2

[x – 2, y +1] = -(2/3)[-4 –x, 5-y] [x – 2 y + 1] = - (2/3)[-4-2 5+1]

3(x- 2) = -2(-4-x) 3(y +1) = -2(5- y) 3(x -2) = -2(-6) 3(y + 1) = -2(6)

3x -6 = 8 +2x 3y + 3 = -10 +2y 3x - 6 = 12 3y + 3 = - 12

X = 14 y = -13 3x = 18 3y = - 15

P(14, -13) x = 6 y = -5

Q (6,-5)

PROBLEM 147. ONE EXTREM OF ONE SEGMENT IS A (2, -5) AND ONE POINT, WITH DISTANCE (1/4) OF THE DISTANCE UNTIL THE ANOTHER EXTREM OF THE SAME SEGMENT, IS B (-1, 4). FIND THE COORDINATES THIS ANOTHER SEGMENT.

AP = (4)AB

[x – 2, y +5] = (4)[-1 -2, 4+5]

(x -2) = -12 (y +5) = 36

x - 2 = -12 y = 31

X = - 10 y = 31 P(-10, 31)

STRAIGTH LINE SLOPE

PROBLEM 157. FIND THE SLOPES OF THE SIDES OF ONE TRIANGLE DETERMINED BY THE POINTS A (-1, 2), B (1,3) AND C(2, -4).

mAB = (3 - 2)/(1 +1) = 1/2 m AC = (-4 - 2)/(2 + 1) = -2 m BC = (-4 -3)/(2 -1) = -7

y – 2 = (1/ 2)(x + 1) y – 3 = -2 (x – 1) y + 4 = -7(x -2)

y = (x/ 2) + 5/2 y = -2x + 5 y = -7x + 10

MATLAB PROGRAM

```
%y = (x/ 2) + 5/2, y = -2x + 5, y = -7x + 10
x = - 5: .001: 5;
y1 = (x/ 2) + 5/2,
y2 = -2*x + 5,
y3 = -7*x + 10,
plot(x, y1, x, y2, x, y3),
xlabel('X');
ylabel('Y');
title('y = (x/ 2) + 5/2, y = -2x + 5, y = -7x + 10 ');
```

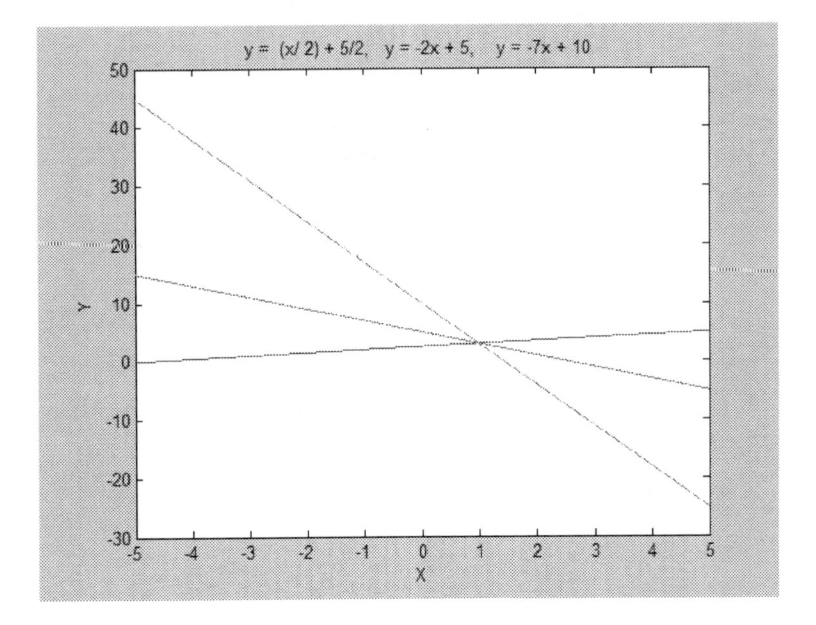

PROBLEM 158. FIND THE ANGLE FORMED BY THE X AXLE AND THE STRAIGTH OF THE POINTS A(1, 1) AND (- 4,5).

m AB = (5 – 1)/(-4 – 1) = -4/5

y – 1 = -(4/ 5)(x – 1)

y = - 4x/5 + 9/5

tg A = (- 4/5 - 0)/(1 – 0) = - 4/5

A = ang tan (-.8)

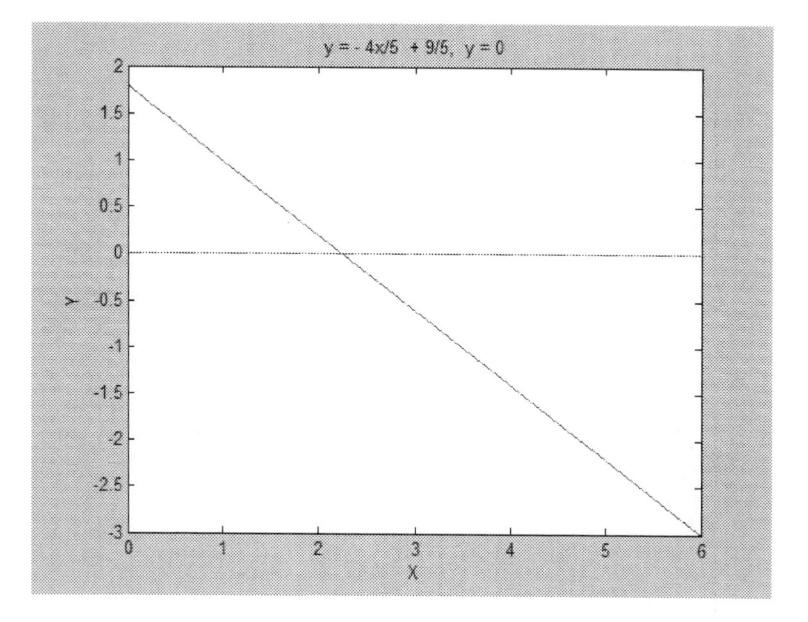

MATLAB PROGRAM

```
%y = - 4x/5 + 9/5
x = -0: .001: 6;
y = - 4*x/5 + 9/5,
y1 = 0,
plot(x, y, x, y1),
xlabel('X');
ylabel('Y');
title('y = - 4x/5 + 9/5, y = 0');
```

PROBLEM 159. THE ANGLES FORMED BY THE AXLE X AND THREE STRAIGTH LINES ARE 45° 120° AND -30° RESPECTIVELY. FIND THE SLOPE OF THE THREE STRAIGTH LINES.

m1 = 1/1 = 1, m2 = - (3)^.5 AND m3 = - 1/((3)^.5)

y 1= x,   y2 = -((3)^^.5)(x - 1)   Y3 = - (1/((3^^.5))) (x – 2)

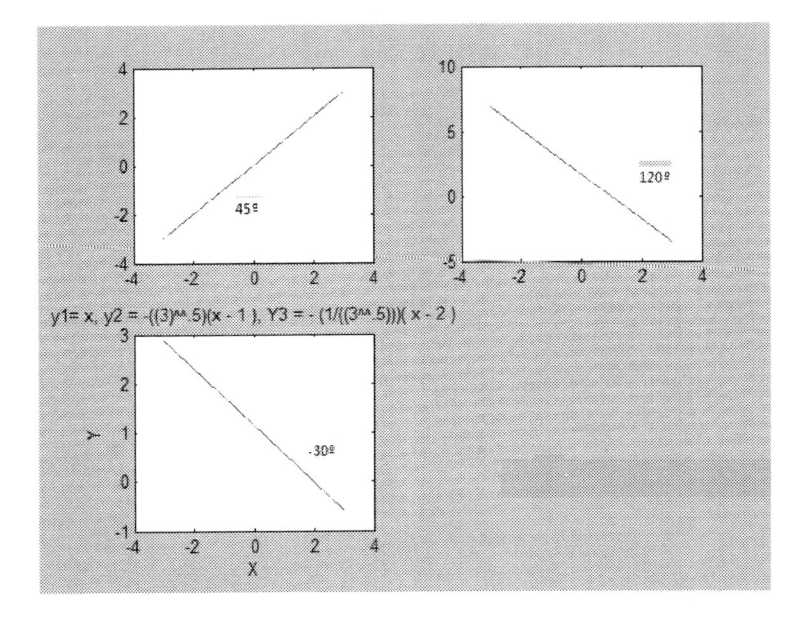

MATLAB PROGRAM

```
%y1= x, y2 = -((3)^^.5)(x - 1), Y3 = - (1/((3^^.5))) (x - 2)
x = -3:   .001:        3;
y1= x,
y2 = -((3)^.5)*(x - 1),
y3 = - (1/((3^.5)))*(x - 2),
subplot(2,2,1), plot(x, y1)
subplot(2,2,2), plot(x, y2)
subplot(2,2,3), plot(x, y3);
xlabel('X');
ylabel('Y');
title('y1= x, y2 = -((3)^^.5)(x - 1), Y3 = - (1/((3^^.5)))(x - 2)');
```

PROBLEM 160. THE SLOPES OF THE SIDES OF ONE TRIANGLE ARE EQUALS m1= 1/ 2, m2 = 1 AND m3 = 2. DEMOSTRATE THAT THE TRIANGLE IS ISOCEL.

tgA =(m2 –m1/(1+ m1m2) = (1 – 1/ 2)/(1 + 1/ 2)= (1/ 2)/(3/ 2) = 1/3,

tg B = (m1 – m3)/(1 + m1m3) = (1/ 2 – 2)/(1 + 1) = - (3/ 2)/(2)= - 3/ 4

tg C = (m3 – m2)/(1 + m2m3) = (2 – 1)/(1 + 2) = 1 /3 THE TRIANGLE IS ISOCEL C = A

PROBLEM 161. DEMOSTRATE THE POINTS A (4, 1), B (1, 2) AND C (-5.4) ARE SITUATED OVER ONE STRAIGTH LINE.

mAB = (2 – 1)/(1 – 4) = - 1/ 3, mBC = (4 – 2)/(-5 – 1) =- 2/6 = -1/3 HAVE THE SAME SLOPE.

Y – 1 = - (1/ 3)(x – 4),

y = 1 – (1/ 3)(x – 4)

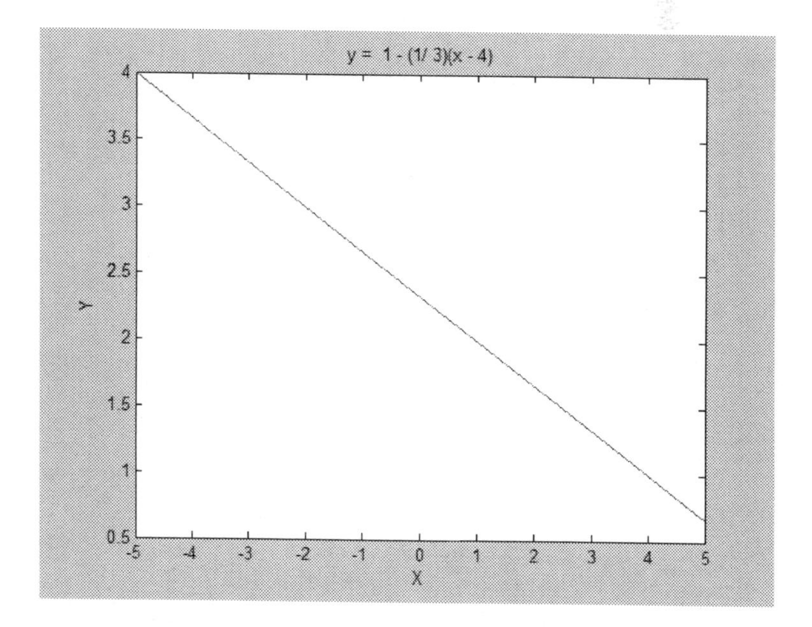

MATLAB PROGRAM

```
%y = 1 - (1/ 3)(x - 4)
x = -5: .001: 5;
y = 1 - (1/ 3)*(x - 4),
plot(x, y);
```

xlabel('X');

ylabel('Y');

title('y = 1 - (1/ 3)(x - 4)');

PROBLEM 162. DEMOSTRATE THE STRAIGTH LINE PASS A (1,-3) AND B (-1, 3) IS PARALLEL AT PASSES BY C (3, -5) AND D (0, 4).

mAB = (-3 -3)/ (1 +1) =- 3 mCD = (4 + 5)/(0 – 3) = -3

y1 + 3 = -3(x -1)

y1 = -3 – 3(x -1)

y2 -3 = -3(x +5)

y2 = 3 – 3(x + 5);

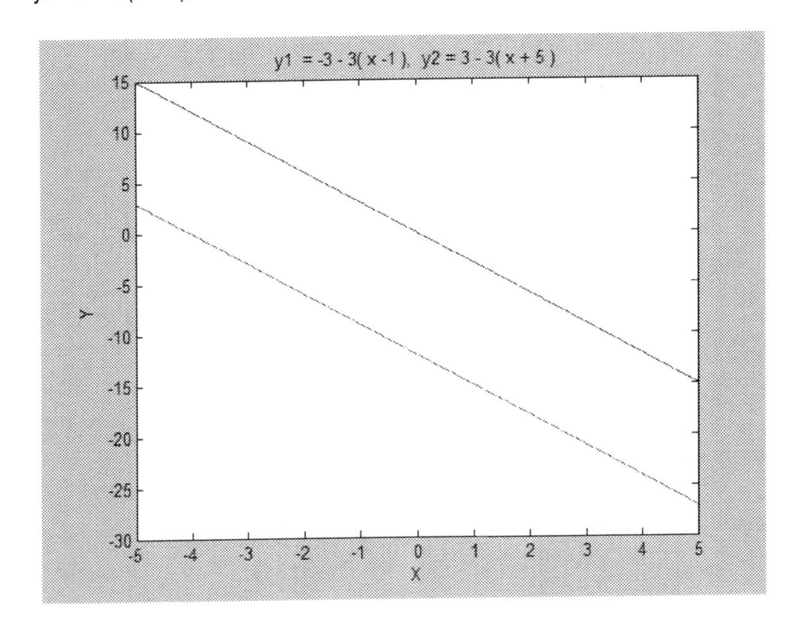


```
%y1 = -3 - 3(x -1), y2 = 3 - 3(x + 5);
x = -5: .001: 5;
y1 = -3 - 3*(x -1),
y2 = 3 - 3*(x + 5);
plot(x, y1, x, y2);
```

```
xlabel('X');

ylabel('Y');

title('y1 = -3 - 3(x -1), y2 = 3 - 3(x + 5)');
```

PROBLEM 163. DEMOSTRATE THE STRAIGTH LINES, DETERMINED BY THE PAIRS OF POINTS A(2,3), B(3,5), AND C (1, 7), D(3,6) ARE PERPENDICULARS.

mAB = (5 – 3)/(3 – 2) = 2, mCD = (6 – 7)/(3 – 1) = - 1/ 2 THE STRAIGTH LINES ARE RECIPROCAL AND CONTRARY SIGN, ARE PERPENDICULAR.

Y1 – 3 = 2(x -2),

y1 = 3 + 2(x – 2),

y2 – 7 = - (1/ 2)(x – 1),

y2 = 7 – (1/2)(x – 1)

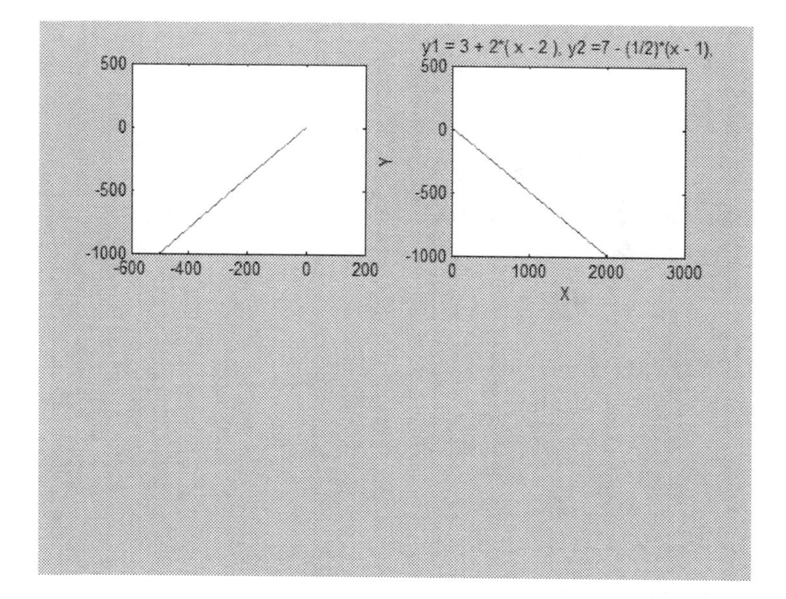


```
%y1 = 3 + 2(x – 2), y2 = 7 - (1/2)(x – 1)

y = -1000: .001: 4;

x1 = 2 + (y - 3)/2 ,
```

```
x2 = 1 -2*(y - 7),
subplot(2,2,1), plot(x1, y);
subplot(2,2,2), plot(x2, y);
xlabel('X');
ylabel('Y');
title('y1 = 3 + 2*(x - 2), y2 =7 - (1/2)*(x - 1),');
```

PROBLEM 165. FIND THE INTERN ANGLES OF A TRIANGLE WITH VERTEX THE POINTS A(2,3), B(-4, 5) AND C(1, -2).

mAB = (5 – 3)/(-4 -2) = -1/ 3 = m1   y – 3 = -(1/ 3)(x – 2)

mAC = (-2 – 3)/(1 -2) = 5 = m2   (y – 3) = 5(x – 2)

mBC = (-2 – 5)/(1 + 4) = -7/ 5 = m3   y +2 = - (7/ 5)(x – 1)

tg A = (m2 – m1)/(1 + m1m2) = (5 + 1/ 3)/(1 – 5/ 3) = (16/3)/(- 2/3) = - 8

tg B = (m3 – m1)/(1 + m1m3) = (-7/ 5 + 1/3)/(1 + (1/ 3)(7/ 5)) =

((-21 + 5)/3)/((1 + 7/15)) = (-16/15)/(22/15) = - 8/11

tg C = (m3 – m2)/(1 + m3m2) = (- 7/ 5 - 5)/(1 - 7) = 32/30 = 16/15

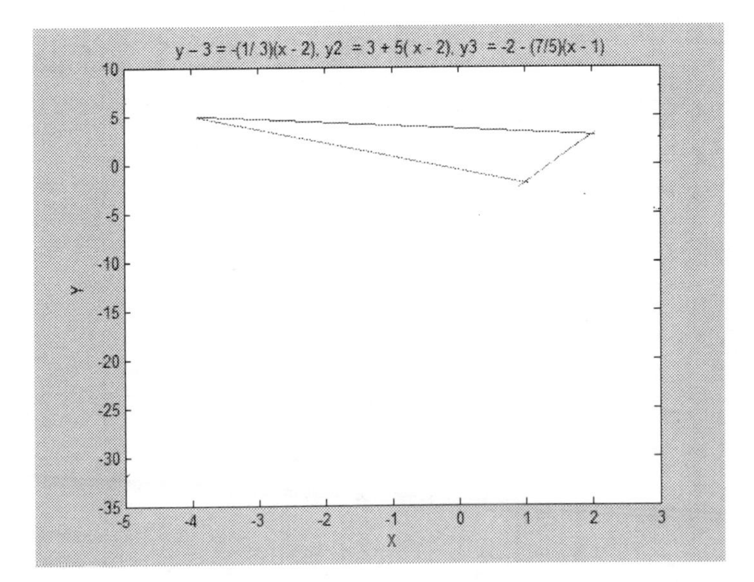

PROBLEM 166. THE POINT E (-2, 2), IS JOINED WITH STRAIGTH LINES AT THE POINTS A(-3, 1), B(0, 2) AND C(1, -1) DEMOSTRATE ONE OF THIS LINES IS BISECT THE ANGLE FORMED BY THE OTHER LINES.

m AE = (2 – 1)/(-2 + 3) = 1 = m1, y – 2 = (x +2) y = 2 + (x + 2)

mBE = (2 – 2)/(-2 – 0) = 0 = m2, y – 2 = 0(x +-2), y = 2,

mCE = (2 + 1)/(-2 – 1) = - 1 = m3, y – 2 = - (x + 2), y = 2 – (x + 2)

tg A = (m1-m2)/(1 + m1m2) = (1 – 0/(1 + 0) = 1

A = ang tg 1 = 45°

tg B = (m2 – m3)/ (1 + m2m3) = (0 + 1)/(1 + 0) = 1

A = ang tg(1) = 45°

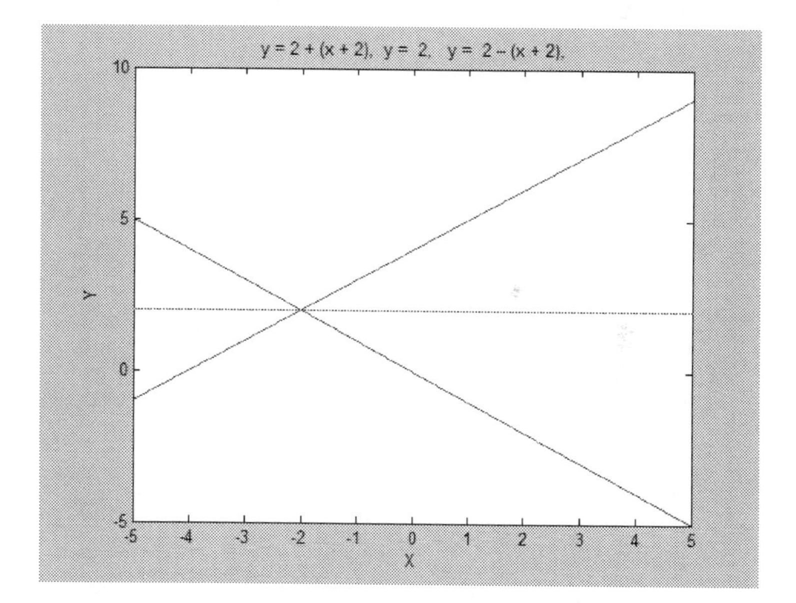

%y = 2 + (x + 2), y = 2, y = 2 – (x + 2),

x = -5: .001: 5;

y1 = 2 + (x + 2),

y2 = 2,

y3 = 2 - (x + 2),

```
plot(x, y1, x, y2, x, y3);
xlabel('X');
ylabel('Y');
title('y = 2 + (x + 2), y = 2, y = 2 – (x + 2),');
```

PROBLEM 167. THE SLOPES OF TWO STRAIGTH LINES ARE EQUALS TO m1 = 2 AND m2 = -3. FIND THE SLOPES OF THE TWO BISECTORS.

mX = BISECTOR SLOPE.

Tg A = (m1 – mX)/(1 + m1mX) = (2 – mX)/(1 + 2mX)

tg B = (mX – m3)/ (1 + mXm3) = (mX + 3)/ (1 + 3mX)

tgA = tg B

(2 – mX)/(1 + 2mX) = (mX + 3)/ (1 - 3mX)

(2 – mX) (1 - 3mX) = (mX + 3) (1 + 2mX)

2 - 6mX – mX + 3mX^2 = mX + 2 mX^2 + 3 + 6mX

mX^2 -14mX - 1 = 0

mX = (14 +- (196 + 4))/2 = 7 +- 5(2)^.5

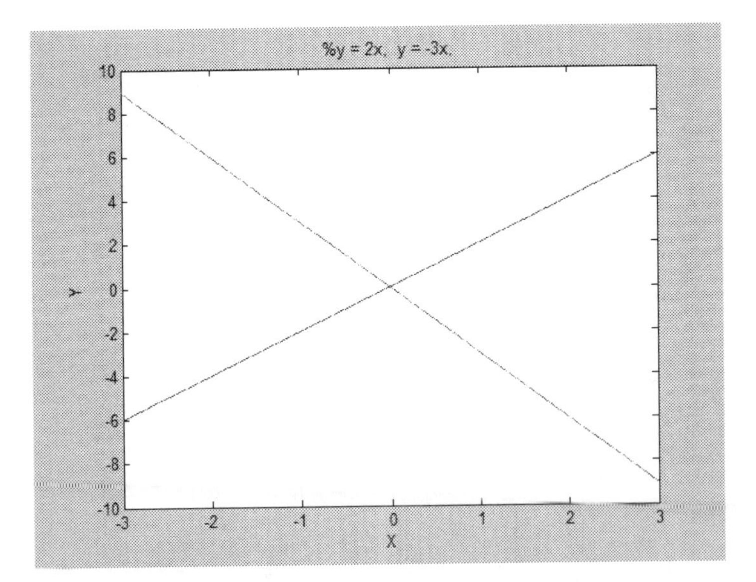

MATLAB PROGRAM

```
%y = 2x, y = -3x,
x = -3: .001: 3;
y1 = 2*x,
y2 = -3*x,
plot(x, y1, x, y2);
xlabel('X');
ylabel('Y');
title('%y = 2x, y = -3x,');
```

PROBLEM 168. FIND THE ANGLE FORMED BY THE STRAIGTH LINES JOIN THE ORIGIN WITH THE POINTS OF TRISECTION OF THE SEGMENT A (-2, 3) AND B (1, -7).

m1 = - 3/2, m2 = - 7.

AP = PB

[x +2 y - 3] = [- 7 – x 1 - y]

x + 2 = -7 – x   y – 3 = 1 - y

2x = -9   2y = 4

X = -9/ 2   y = 2

P(-9/2, 2)   m3 = 2/(- 9/2) = - 4/9

tg A = (m1 – m2)/(1 + m1m2) = (-3/2 + 7)/(1 + 21/2))

= (-3/2 + 7)/(1 + 21/2) = ((11)/2)/(23/2) = (11)/(23) =

PROBLEM 169. THE SLOPE OF AB IS 1/ 2. FIND THE SLOPE OF CD IF THE ANGLE FORMED BY AB AND CD IS OF 30ª.

Tg 30º = 1/(3)^.5 = (1/ 2 – m2)/(1 + m2/2)

(1 + m2/2) = (3)^.5(1/ 2 – m2)

m2(1/2 + (3)^.5) = ((3)^.5)/2 -1)

m2(1 + 2(3)^^.5)/2 = ((3)^.5 - 2)/2

m2 (((3)^.5)2 +1) = (-2 + (3)^.5)

m2 = (((3)^.5)2 -1) (-2 + (3)^.5)/ (((3)^.5)2 +1) (((3)^.5)2 -1) =

(-5(3)^^.5 + 6 + 2)/(12 – 1) =

8/11 – (5(3)^^.5) /11

PROBLEM 170. DEMOSTRATE THE POINTS A(6, 11), B(-11,4), C(-4, 13) AND D (1, -14) IS SITUATED OVER ONE CIRCUMFERENCE.

AP1 = P1B [x – 6, y - 11] = [-11 –x, 4 - y]   m AB = (4 – 11)/(-11 -6) = 7/17

X – 6 = -11 – x   y - 11 = 4 -y   PERPENDICULAR SLOPE - 17/ 7

2x = -5   2y = 15   y – 15/2 = -(17/7)(x + 5/ 2)

X = -5/ 2   y = 15/ 2   y = 15/ 2 - (17/7)(x + 5/ 2)

CP2 = P2D [x + 4, y - 13] = [1- x, -14 - y]   m CD =

X + 4 = 1 – x   y – 13 = -14 – y

2x = - 3   2y = - 1

x = - 3/ 2   y = -1/ 2

STRAIGTH LINE

PROBLEM 212. SUPPOSE THE ANGLE FORMED BY THE AXLE X AND STRAIGTH LINE IS 60ª, AND THE STRAIGTH LINE PASS (-1. -3). DETERMINE IT`S EQUATION.

Tg 60ª = ((3)^.5) = slope

y – y1 = m(x – x1)

y +3 = ((3)^.5)(x + 1)

y = -3 + ((3)^.5)(x + 1)

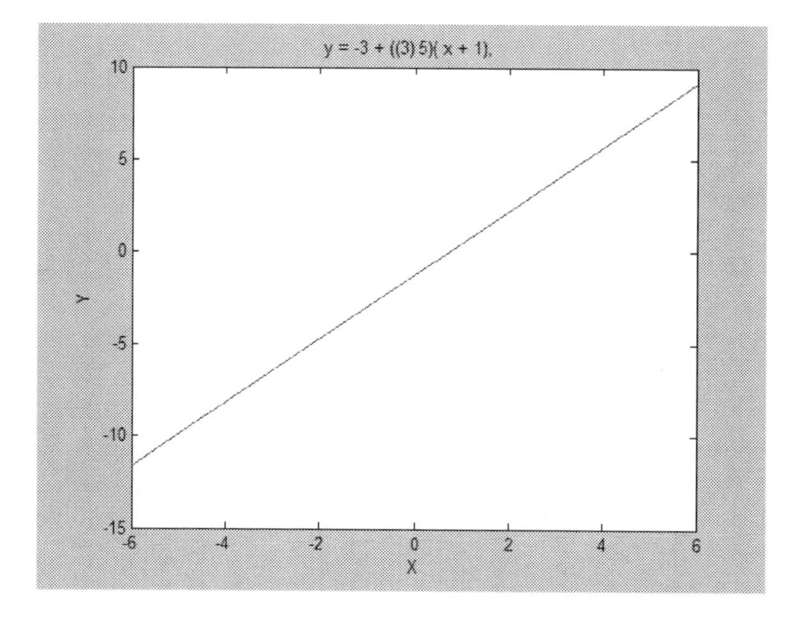

%y = -3 + ((3)^.5)(x + 1),

x = -6: .001: 6;

```
y = -3 + ((3)^.5)*(x + 1),
plot(x, y);
xlabel('X');
ylabel('Y');
title('y = -3 + ((3)^.5)(x + 1), ');
```

PROBLEM 213. FIND THE EQUATION OF THE STRAIGTH LINE PASS BY (2, -1) AND (3, 2).

m = (2 + 1)/(3 – 2) = 3

y – y1 = m (x – x1)

y + 1 = 3(x – 2)

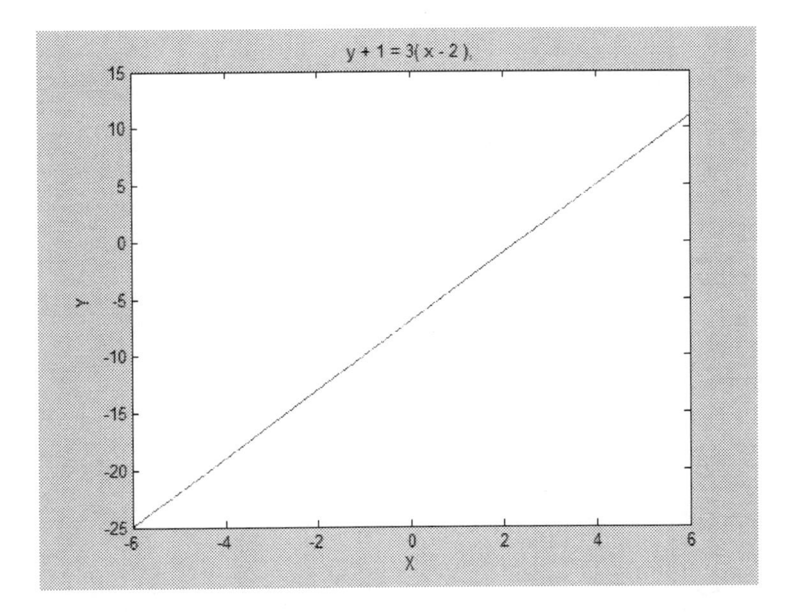


```
%y + 1 = 3(x - 2),
x = -6: .001: 6;
y = -1 + 3*(x - 2),
plot(x, y);
xlabel('X');
```

```
ylabel('Y');
title(' y + 1 = 3(x - 2), ');
```

PROBLEM 214. FIND THE EQUATION OF THE STRAIGTH LINE PASS BY (2, 3) AND (2, -4).

m = (-4 -3)/(2 – 2) = infinito

y – y1 = m (x – x1) ,

x – x1 = (y – y1)/infinito = 0

x = 2,

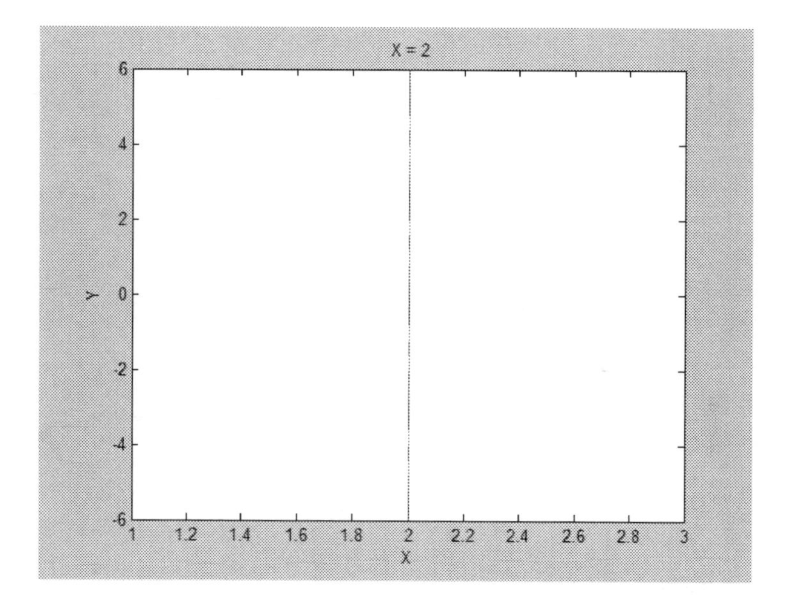


```
%y + 1 = 3(x - 2),
y = -6: .001: 6;
x = 2;
plot(x, y);
xlabel('X');
ylabel('Y');
title(' X = 2 ');
```

PROBLEM 215. FIND THE EQUATION OF THE STRAIGTH LINE PASS BY (1, 2) AND IS PARALELL AT THE X AXLE.

Y = 2, m = 0,

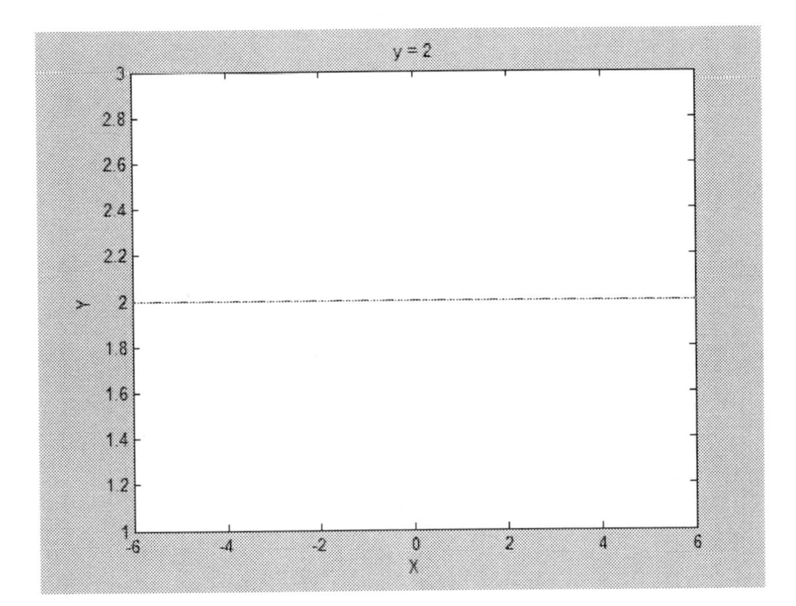


```
%y + 1 = 3(x - 2),
x = -6: .001: 6;
y = 2;
plot(x, y);
xlabel('X');
ylabel('Y');
title(' y = 2 ');
```

PROBLEM 216. FIND THE EQUATION OF THE PERPENDICULAR IN THE MIDDLE POINT OF THE SEGMENT JOIN A(-3, 5) AND B(-4,1).

AP = PB [x + 3, y - 5] = [-4 –x, 1 -y]

x + 3 = -4 – x y – 5 = 1 – y

2x = -7 2y = 6 x = -7/ 2 y = 3 P(-7/ 2, 3)

PROBLEM 217. ONE EQUILATERAL TRIANGLE HAS IT'S BASE IN THE AXLE X, AND ITS VERTEX IN (3, 5) FIND THE EQUATIONS OF IT'S SIDES.

m1 = tg 60º = (3)^.5   m2 = tg -60ª = - (3)^.5

y – 5 = (3)^.5(x – 3)

y – 5 = -(3)^.5(x – 3)

PROBLEM 218. ONE STRAIGTH LINE IS PERPENDICULAR AT THE SEGMENT A (-4,-2) AND B (2, -6) IN THE POINT DISTANT THE THIRD PART OF THE LONGITUD OF THE SEGMENT. DETERMINE its EQUATION.

AP = (1/ 3)AB

[x + 4, y + 2] = (1/ 3) [2 +4, 2 - 6]   mAB = (-2 + 6)/(-4 -2) = - 4/6 = - 2/ 3

X + 4 = 2   y + 2 = - (1/ 3)4   PERPENDICULAR SLOPE 3/ 2

X = -2   y = -10/3   P(-2 , -10/ 3)

y + 10/3 = (3/ 2)(x + 2) 3y + 10 = 9x/2 + 18

6y + 20 = 9x + 18

6y - 9x = -2

PROBLEM 219. FIND THE EQUATION OF THE STRAIGTH LINE PASS BY A (3, 5) AND IS PARALLEL AT THE LINE PASS B (2, 5) AND C (-5,-2).

mBC = (-2 – 5)/(-5 – 2) = 1

y – 5 = 1(x – 3)

y – x -2 = 0

PROBLEM 220. THE DIAGONAL ONE PARALELLOGRAM JOIN THE VERTEX A (4,-2) AND B (-4,-4). THE EXTREM OF THE ANOTHER DIAGONAL IS C (1, 2) FIND THE EQUATION AND LONGITUDE.

THE MIDLE POINT OF AB IS P

AP = PB

[x – 4, y + 2] = [-4 – x, -4 - y]

X – 4 = -4 –x   y + 2 = -4 -y

x = 0   2y = -6   y = -3   P(0, -3)

SLOPE CP = (-3 – 2)/(0 -1) = 5

STRAIGTH LINE Y – 3 = 5(X -0)

LONGITUDE = 2(CP) = 2(1 + 25)^.5 = 2(26)^.5

PROBLEM 221 THE DIAGONAL ONE SQUARE JOIN THE VERTEX A (1, 2) AND B (2, 5).

FIND THE EQUATIONS OF THE SIDES OF THE SQUARE.

SLOPE AB = (2 -5)/(1- 2) = 3

tg 45 = = 1 = (m2 – m1)/(1 + m1m2) = (3 –m1)/(1 + 3m1)

1 + 3m1 = 3 – m1   4m1 = 2

m1 = 1 / 2

-1 -3m1 = 3 –m1

-2m1 = 4

m1 = -2

STRAIGTH LINES POINT A y – 2 = (1/ 2)(X – 1)

2Y – 4 = x - 1

2y –x = 3

y -2 = -2(x -1) y – 2 = -2x -2

y + 2x = 4

STRAIGTH LINES POINT B (2,5) y –5 = (1/ 2)(X – 2)

2Y – 10 = x - 2

2y –x = 8

y -5 = -2(x -2) y – 5 = -2x + 4

PROBLEM 222. THE BASE OF ISOSELES TRIANGLE ARE THE POINTS A (-2, 3) AND B (3, -1). ITS VERTEX IS OVER AXLE Y. FIND THE EQUATIONS OF ITS SIDES.

VERTEX P (0,Y)   DISTANCE AP = PB

(4 + (Y -3)^2)^.5 = (9 + (-1 –Y)^2)^.5

4 + Y^2 – 6Y +9 = 9 + 1 +2Y + Y^2

$13 - 6Y = 10 + 2y$ $8y = 3$

$y = 3/8$ $P(0, 3/8)$

PROBLEM 223. DEMOSTRATE THE EQUATION $2X - Y = 3$ REPRESENT A STRAIGTH LINE FINDS ITS SLOPE.

$Y = 2X - 3$ IT'S A FIRST DEGRE EQUATION WITH SLOPE EQUAL 2

CIRCUMFERENCE

PROBLEM 276. WHAT IS THE ROOT LOCUS REPRESENTED BY THE EQUATION x^2 + y^2 + 2x + +2y +2 = 0.

(x + 1)^2 + (y + 1)^2 + 2 – 2 = 0

(x + 1)^^2 + (y + 1)^2 = 0 THE EQUATION REPRESENT ONE POINT (-1.-1) WITH RADIUS = 0

PROBLEM 277. WHAT'S THE ROOT LOCUS CORRESPONDING AT THE EQUATION x^2 + y^2 – 6x + 6y + 9 = 0?

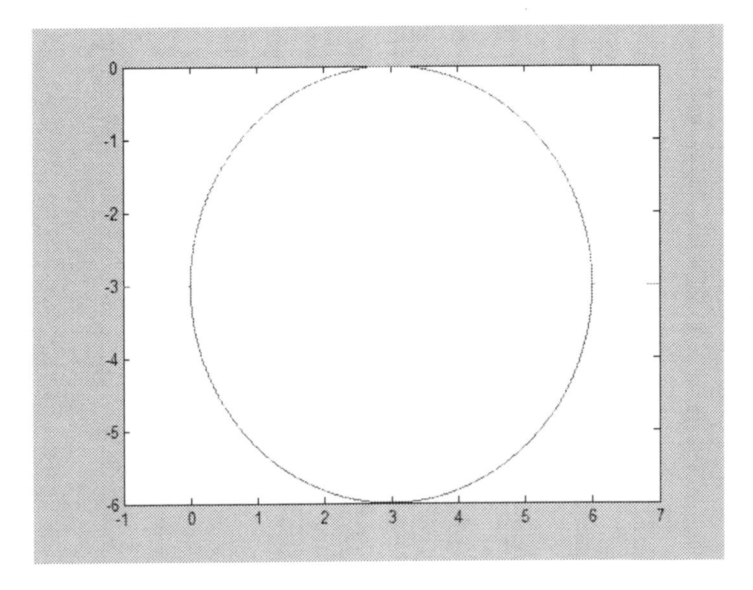

MATLAB SOLUTION

x = -1: .001: 7;

z = (x - 3);

```
real_z1 = z.*z;
z = real_z1;
y2 = -3 + sqrt((9 - z));
y1 = -3 - sqrt((9 - z));
real_y1 = -3 - sqrt((9 - z));
y1 = real_y1;
y2 = -3 + sqrt((9 - z));
real_y2 = -3 + sqrt((9 - z));
y2 = real_y2;
plot (x, y1, x, y2);
```

PROBEM 278. FIND THE EQUATION OF THE CIRCUNFERENCE PASS BY THE POINT (-2, 4) AND HAVE THE SAME CENTER THAT THE REPRESENTED BY THE EQUATION x^2+ y^2-5x + 4y -1 = 0.

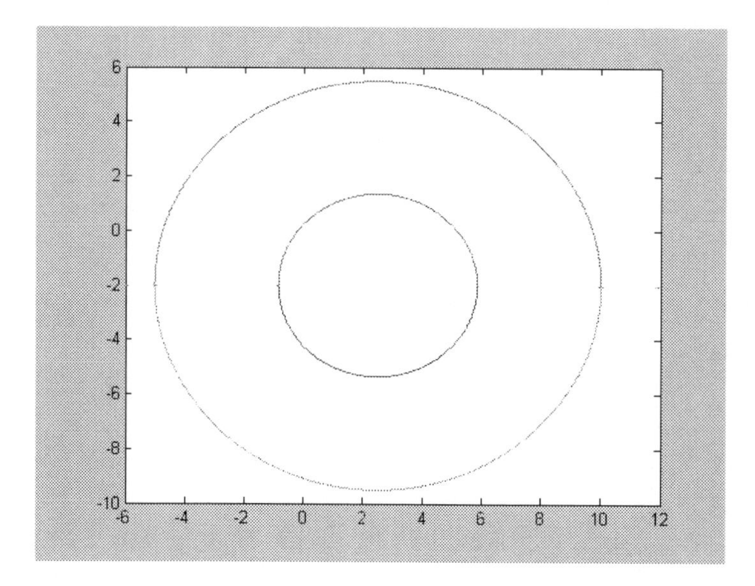

X^2+ y^2 – 5x + 4y – 1 = 0

(x – 5/2)^2 + (y + 2)^2 -1 – 4 – 25/4 = 0 ,

$(x - 5/2)^2 + (y + 2)^2 = 45/4$ CENTER C(5/2, -2),

$R^2 = (5/2 + 2)^2 + (-2 -4)^2 = 81/4 + 36 =(81 + 144)/4 = 225/4$

$(x- 5/2)^2 + (y -2)^2 = 225/4$

MATLAB SOLUTION

```
x = -6: .001: 12;
z = (x - 5/2);
real_z1 = z.*z;
z = real_z1;y2 = -2 + sqrt((45/4 - z));
y1 = -2 - sqrt((45/4 - z));
real_y1 = -2 - sqrt((45/4 - z));
y1 = real_y1;
y2 = -2 + sqrt((45/4 - z));
real_y2 = -2 + sqrt((45/4 - z));
y2 = real_y2;
x = -6: .001: 12
z = (x - 5/2);
real_z1 = z.*z;
z = real_z1;
y3 = -2 + sqrt((225/4 - z));
real_y3= -2 + sqrt((225/4 - z));
y3 = real_y3;
y4 = -2 - sqrt((225/4 - z));
real_y4 = -2 - sqrt((225/4 - z));
y4 = real_y4;
plot (x, y1, x, y2, x, y3, x, y4);
```

PROBLEM 279. FIND THE EQUATION OF THE CIRCUMFERENCE WITH DIAMETER IS THE SEGMENT THAT JOIN (-1, -2) AND (3, 4).

X1 = (-1 +3)/2 = 1, y1 = (-2 +4)/2 = 1 CENTER C(1, 1) ,

R^2 = (1 +1)^2`+ (1 + 2)^2 = 4 + 9 = 13

(x- 1)^2 + (y – 1)^2 = 13

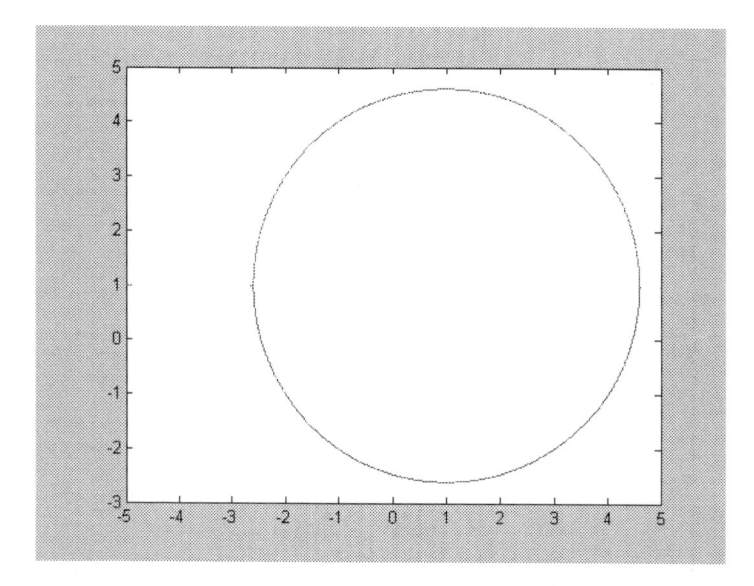

MATLAB SOLUTION

```
x = -5: .001: 5;
z = (x - 1);
real_z1 = z.*z;
z = real_z1;
y1 = 1 - sqrt((13 - z));
real_y1 = 1 - sqrt((13 - z));
y1 = real_y1;
y2 = 1 + sqrt((13 - z));
real_y2 = 1 + sqrt((13 - z));
y2 = real_y2
plot (x, y1, x, y2);
```

PROBLEM 280. FIND THE EQUATION OF THE CIRCUMFERENCE PASS BY (1, 2) AND ARE TANGENTS AT THE COORDINATE AXLES.

PASS BY A(1,2), B(0, y), D(x,0) CENTER (h,k), (y – 2) = (k/h) (x - 1) ,

(1 – h)^2 + (2 – k)^2 = r^2, (E) h = k =r (1 – h)^^2 + (2 – h)^^2 = r^^2

1 – 2h + h^^2 + 4 – 4h + h^^2 = h^`2

h^2- 6h + 5 = 0,

h = (6 +- (36 – 20)^^.5)/2 = 6 +- 4)/2,

h1 = 5 h2 = 1,

k 1 = 5 k2 = 1,

r1 = 1 r2 = 5

(x – 1)^^2 + (y – 1)^^2 = 1

(x – 5)^^2 + (y – 5)^^2 = 25

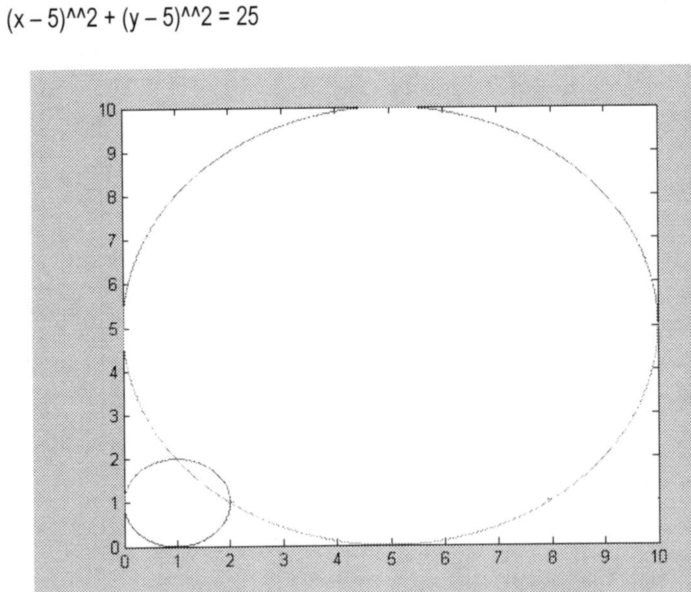

MATLAB SOLUTION

```
x = 0: .001: 10;
z = (x - 1);
real_z1 = z.*z;
z = real_z1;
if ((1 < z))
fprintf('IMAGINARY NUMBER %f /n')
else
y1 = 1 - sqrt((1 - z));
real_y1 = 1 - sqrt((1 - z));
end
y2 = 1 + sqrt((1 - z));
x = 0: .001: 10;
z = (x - 5);
real_z1 = z.*z;
z = real_z1;
if ((25 < z))
fprintf('IMAGINARY NUMBER %f /n')
else
y3 = 5 + sqrt((25 - z));
end
y4 = 5 - sqrt((25 - z));
plot (x, y1, x, y2, x, y3, x, y4)
```

PROBLEM 281.

FIND THE EQUATIONS OF THE CIRCUMFERENCES THAT HAVE ITS CENTERS IN

THE ORIGIN AND ARE TANGENTS AT THE CIRCUMFERENCE x^2 + y^2 – 4x + 4y +7 = 0

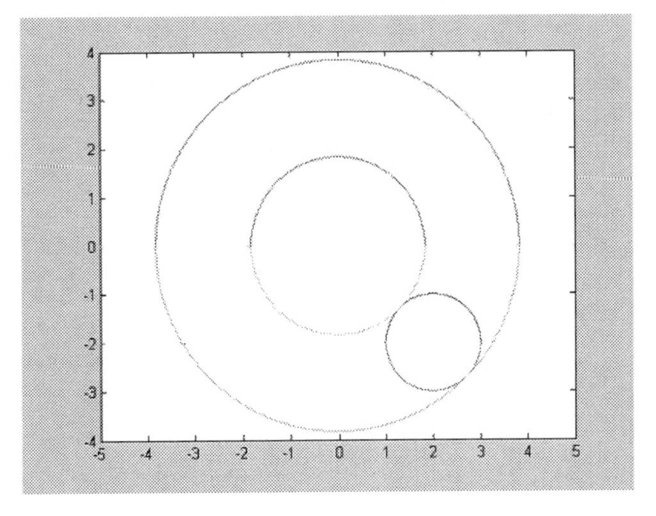

X^2 + y^2 – 4x + 4y +7 = 0

(x -2)^2 + (y + 2)^2 + 7 -8 = 0

(x – 2)^2 + (y + 2)^2 = 1 , CENTER AND RADIUS C(2, -2) R = 1

y – 0 = -1(x – 0),

y = -x,

THE TANGENT POINTS AT THIS CIRCUMFERENCES ARE

(x – 2)^2 + (-x +2)^2 = 1,

x^2 – 4x + 4 + x^2 – 4x + 4 -1= 0,

2x^2 – 8x + 7 = 0,

x = (8 +- (64 – 56)^^.5)/4 = 2 +- ((8)^^.5)/4 y = -2 +- ((8)^.5)/4

r^^2 = (2 +- ((8)^.5)/4)^2 + (-2 +- ((8^^.5)/4)^^2 = 4 +- (8)^.5 + 2/4 + 4 +- (8)^^.5 + 2/4

r^^2 = 9 +- 2(8)^^.5 OF BOTH TANGENTS CIRCUMFERENCE

x^2 + y^2 = r^^2 = 9 +- 2(8)^.5

MATLAB SOLUTION

```
x = -5: .001: 5;
z = (x - 2);
real_z1 = z.*z;
z = real_z1;
y1 = -2 + sqrt((1 - z));
real_y1 = -2 + sqrt((1 - z));
y1 = real_y1;
y = -2 - sqrt((1 - z));
real_y2 = -2 - sqrt((1 - z));
y2 = real_y2;
plot (x, y1, x, y2);
x = -5: .001: 5;
z = (x);
real_z1 = z.*z;
z = real_z1;
y3 = sqrt((9 + 2*(8)^.5 - z));
real_y3= sqrt((9 + 2*(8)^.5 - z));
y3 = real_y3;
y4 = - sqrt((9 + 2*(8)^.5 - z));
real_y4 = - sqrt((9 + 2*(8)^.5 - z));
y4 = real_y4;
plot (x, y1, x, y2, x, y3, x, y4);
x = -5: .001: 5;
z = (x);
real_z1 = z.*z;
z = real_z1;
y5 = sqrt((9 - 2*(8)^.5 - z));
real_y5= sqrt((9 - 2*(8)^.5 - z));
```

```
y5 = real_y5;
y6 = - sqrt((9 - 2*(8)^.5 - z));
real_y6 = - sqrt((9 - 2*(8)^.5 - z));
y6 = real_y6;
plot (x, y1, x, y2, x, y3, x, y4, x, y5, x, y6);
```

PROBLEM 282. FIND THE INTERSECTION OF THE CIRCUMFERENCES

X^2 + y^2 = 2x + 2y (A)

X^2 + y^2 + 2x = 4 (B)

X^2 + y^2 – 2x – 2y = 0 OBTAIN A - B

-x^2 - y^2 – 2x + 4 = 0

- 4x – 2y + 4 = 0, y = - 2x + 2 (C)

C in A x^^2 + (-2x+2)^^2 = 2x + 2y

X^^2 + 4x^^4 - 8x + 4 = 2x + 2y = 2x + (-4x +4) = -2x +4

5x^^2 - 6x = 0

(x)(5x – 6) = 0 x1 = 0 AND x2 = 6/5 y1 = 2 y2 = -12/5 + 10/5 = -2/5

MATLAB SOLUTION

```
x = -6: .001: 12;
z = (x - 5/2);
real_z1 = z.*z;
z = real_z1;y2 = -2 + sqrt((45/4 - z));
y1 = -2 - sqrt((45/4 - z));
real_y1 = -2 - sqrt((45/4 - z));
y1 = real_y1;
y2 = -2 + sqrt((45/4 - z));
real_y2 = -2 + sqrt((45/4 - z));
y2 = real_y2;
x = -6: .001: 12
```

```
z = (x - 5/2);
real_z1 = z.*z;
z = real_z1;
y3 = -2 + sqrt((225/4 - z));
real_y3= -2 + sqrt((225/4 - z));
y3 = real_y3;
y4 = -2 - sqrt((225/4 - z));
real_y4 = -2 - sqrt((225/4 - z));
y4 = real_y4;
plot (x, y1, x, y2, x, y3, x, y4);
```

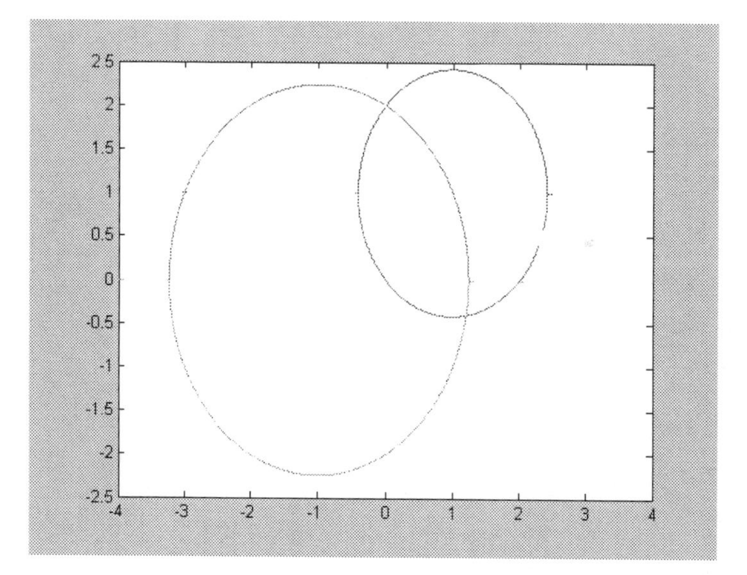

PROBLEM 283. FIND THE EQUATION OF THE CIRCUMFERENCE PASS BY THE THREE POINTS (0,3), (3,0) AND (0,0).

C (3/2,3/2) R^2 = (0- 3/2)^2 + (0 – 3/2)^2 = 9/4 + 9/4 = 9/2

(x – 3/2)^^2 + (y – 3/2)^^2 = 9/2 , x^^2 - 3x + 9/4 + y^^2 – 3y + 9/4 = 9/2

X^2 + y^2 – 3x - 3y = 0

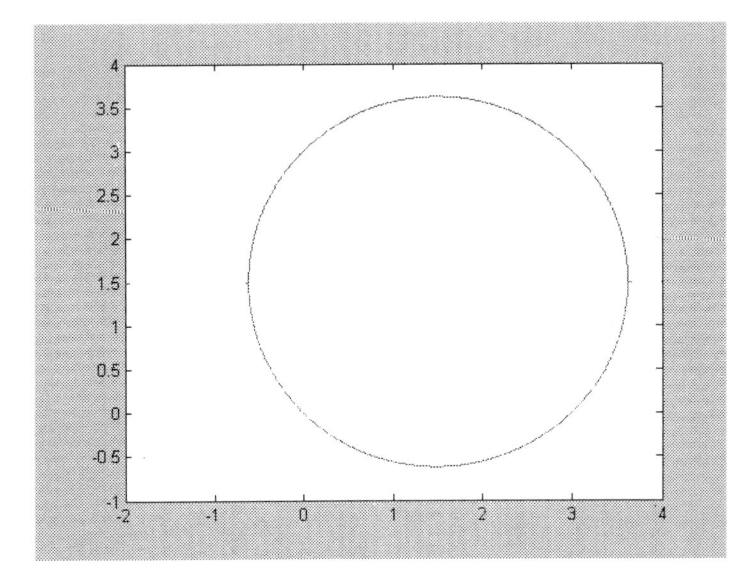

MATLAB SOLUTION

```
x = -2: .001: 4;
z = (x - 3/2);
real_z1 = z.*z;
z = real_z1;
y1 = 3/2 + sqrt((9/2 - z));
real_y1 = 3/2 + sqrt((9/2 - z));
y1 = real_y1;
y = 3/2 - sqrt((9/2 - z));
real_y2 = 3/2 - sqrt((9/2 - z));
y2 = real_y2;
plot (x, y1, x, y2);
```

PROBLEM 284. FIND THE CIRCUMFERENCES OF RARIUS 5 AND PASS BY THE POINTS (2,-1) AND (3,-2).

(2 –h)^2 + (-1 – k)^2 = 25

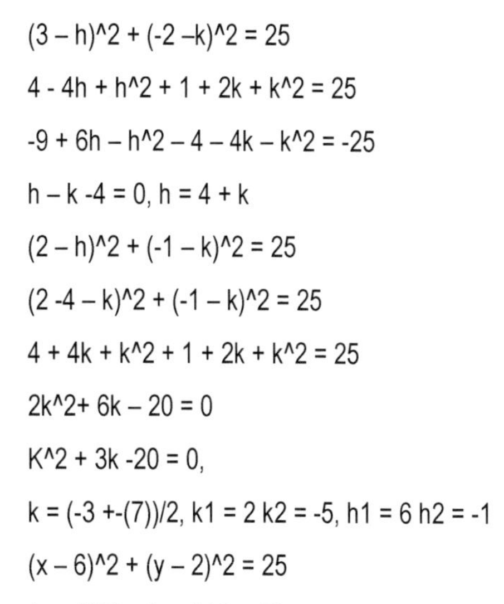

(3 – h)^2 + (-2 –k)^2 = 25

4 - 4h + h^2 + 1 + 2k + k^2 = 25

-9 + 6h – h^2 – 4 – 4k – k^2 = -25

h – k -4 = 0, h = 4 + k

(2 – h)^2 + (-1 – k)^2 = 25

(2 -4 – k)^2 + (-1 – k)^2 = 25

4 + 4k + k^2 + 1 + 2k + k^2 = 25

2k^2+ 6k – 20 = 0

K^2 + 3k -20 = 0,

k = (-3 +-(7))/2, k1 = 2 k2 = -5, h1 = 6 h2 = -1

(x – 6)^2 + (y – 2)^2 = 25

(x + 5)^2 + (y +1)^2 = 25

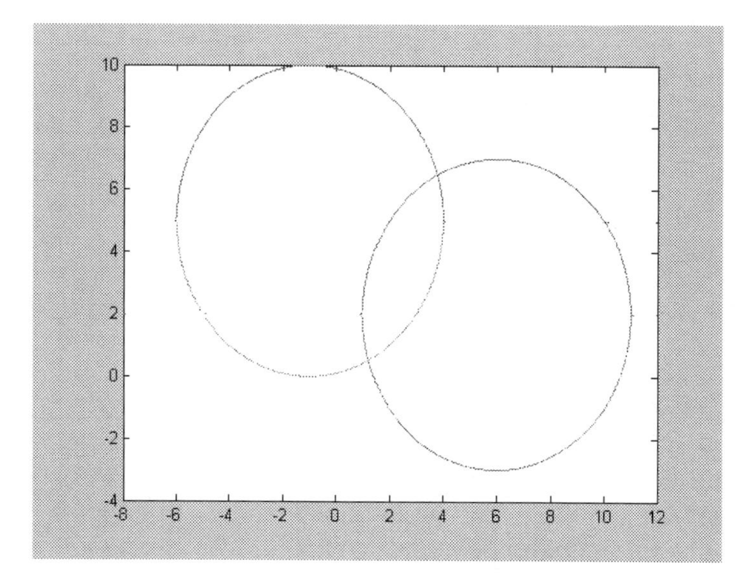

MATLAB SOLUTION

%(x -17/18)^2 + (y - 17/18)^2 = r^2

%(y - 71/18)^2 = r^2 - (x - 17/18)^2

```
x = -8: .001: 12;
z = (x - 6);
real_z1 = z.*z;
z = real_z1;
y1 = 2 + sqrt((25 - z));
real_y1 = 2 + sqrt((25 - z));
y1 = real_y1;
y = 2 - sqrt((25 - z));
real_y2 = 2 - sqrt((25 - z));
y2 = real_y2;
x = -8: .001: 12;
z = (x + 1);
real_z1 = z.*z;
z = real_z1;
y3 = 5 + sqrt((25 - z));
real_y3 = 5 + sqrt((25 - z));
y4 = real_y3;
y3 = 5 - sqrt((25 - z));
real_y4 = 5 - sqrt((25 - z));
y2 = real_y2;
plot (x, y3, x, y1, x, y2, x, y3, x, y4);
```

PROBLEM 285. THE CENTER OF ONE CIRCUMFERENCE PASS BY THE POINTS (1, -2) AND (-2, 2) IS OVER THE STRAIGTH LINE $8x - 4y + 9 = 0$ WHICH IS IT'S EQUATION?

```
x = -5: .001: 2.8;
y3 = 2*x + 9/4;
x = - 5: .001: 2.8;
%(x -17/18)^2 + (y - 17/18)^2 = r^2
```

```
%(y - 71/18)^2 = r^2 - (x - 17/18)^2

z = (x + 3/2);

real_z1 = z.*z;

z = real_z1;

y1 = -3/4 + sqrt((125/16 - z));

real_y1 = -3/4 + sqrt((125/16 - z));

y1 = real_y1;

y = -3/4 - sqrt((25/16 - z));

real_y2 = -3/4 - sqrt((125/16 - z));

y2 = real_y2;

plot (x, y3, x, y1, x, y2);
```

PROBLEM 286.FIND THE EQUATION OF THE CIRCUMFERENCE PASS BY THE POINTS (0,0) AND (2, -2) AND IS TANGENT AT THE STRAIGTH LINE y + 4 = 0.

(x –h)^2 + (y –k)^2 = r^2

h^2 + y^2 = r^2

(2 – h)^2 + (-2 –k)^2 = r^2

4 – 4h + h^2 + 4 + 4k + k^2 = r^2

8- 4h +4k + h^2 + k^2 = r^2

-h^2 -k^2 = -r^2

8 – 4h + 4k

k = h – 2 y = -4

h^2 + k^2 = r^^2 = (k + 4)^2 = k^2 + 8k + 16

h^2 - 8k -16 =0

h^2 - 8(h – 2) -16 = 0

h^2 -8h +16 – 16 = 0

h^2 - 8h = 0 h1 = 0, h2 = 8 k1 = -2 k2 = 6

(x – 0)^2 + (y +2)^2 = 4

(x- 8)^^2 + (y- 6)^^2 = 64 + 36 = 100

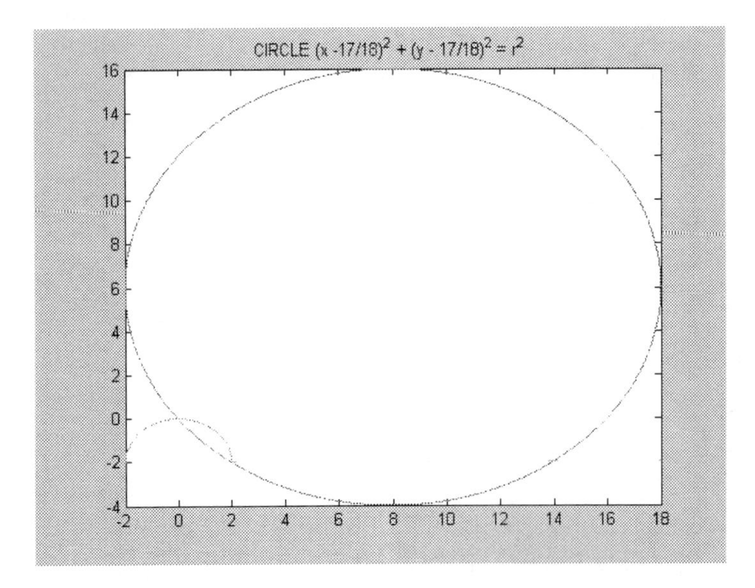


```
x = -2: .001: 18;
z = (x);
real_z1 = z.*z;
z = real_z1;
y1 = -2 + sqrt((4 - z));
real_y1 = -2 + sqrt((4 - z));
y1 = real_y1;
y = -2 - sqrt((4 - z));
real_y2 = -2 - sqrt((4 - z));
y2 = real_y2;
x = -2: .001: 18;
z = (x - 8);
real_z1 = z.*z;
z = real_z1;
```

```
y3 = 6 + sqrt((100 - z));
real_y3 = 6 + sqrt((100 - z));
y3 = real_y3;
y4 = 6 - sqrt((100 - z));
real_y2 = 6 - sqrt((100 - z));
y2 = real_y2;
plot (x, y2, x, y3, x, y4, x, y1);
title('CIRCLE (x -17/18)^2 + (y - 17/18)^2 = r^2 ');
```

PROBLEM 287. FIND THE EQUATION OF THE CIRCUMFERENCE PASS BY THE POINTS A(-1,0) AND B(0,1) AND IS TANGENT AT THE AT THE STRAIGTH LINE x – y = 1.

AP = PB   mAB = (1 – 0)/(0 +1) = 1

[x +1 y -0] = [0 –x, 1 - y]   PERPENDICULAR SLOPE = -1

x + 1 = - x   y = 1 – y

x = - 1/ 2   y = 1/ 2 P(-1/ 2, 1/ 2)   k – 1/ 2 = (-1) (h + 1/ 2) =- h – 1/ 2

h = -k r = (h – k – 1)/(2)^.5 = (2h -1)/(2)^.5 = (-2k -1)/(2)^^.5

(x – h)^2 + (y – k)^2 = r^2 = ((2h – 1)^2)/2

(-1 –h)^2 + k^2 = (4h^2 – 4h + 1)/2

2 + 4h + 2h^2 + 2k^2 = 4h^2 – 4h + 1

8h = 1 h = -1/8 k = 1/8 C(-1/8, 1/8)

(x + 1/8)^2 + (y – 1/8)^2 = (5/(4(2)^.5))^^2 = 25/32

(x^2 + x/4 + 2/64 + y^2 - y/4) = 25/32

x^2 + y^2 + x/4 – y/4 – 24/32 = 0

4 x^2 + 4y^2 + x – y - 3 = 0

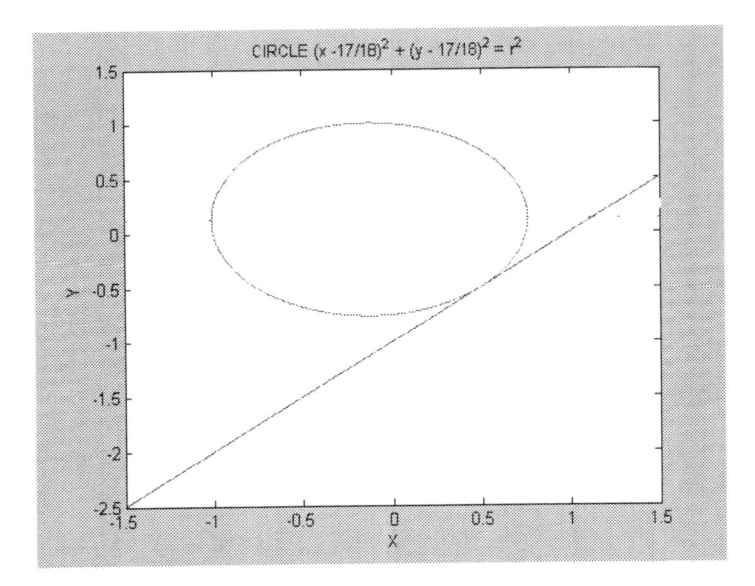


```
x = -1.5: .001: 1.5;
y3 = x - 1;
x = - 1.5: .001: 1.5;
%(x -17/18)^2 + (y - 17/18)^2 = r^2
%(y - 71/18)^2 = r^2 - (x - 17/18)^2
z = (x + 1/8);
real_z1 = z.*z;
z = real_z1;
y1 = 1/8 + sqrt((25/32 - z));
real_y1 = 1/8 + sqrt((25/32 - z));
y1 = real_y1;
y = 1/8 - sqrt((25/32 - z));
real_y2 = 1/8 - sqrt((25/32 - z));
y2 = real_y2;
plot (x, y3, x, y1, x, y2);
```

```
xlabel('X');
ylabel('Y');
title('CIRCLE (x -17/18)^2 + (y - 17/18)^2 = r^2 ');
```

PROBLEM 288. WHAT'S THE EQUATION OF THE CIRCUMFERENCE IS TANGENT AT THE STRAIGTH LINES x = 3 AND x = 7 WITH CENTER IS OVER THE STRAIGTH LINE y = 2x + 4?.

h = (3 + 7)/2 = 5 r = (7 -3)/2 =2 k = 2h + 4

K = 10 + 4 = 14 C(5, 14)

(x - 5)^^2 + (y – 14)^^2 = 4

X^^2 - 10x + 25 + y^^2 – 28y + 196 = 25

X^2 + y^2 – 10x – 28y + 196 = 0

MATLAB SOLUTION

```
x = 3: .001: 7;
y3 = 2*x + 4;
x = 3: .001: 7;
%(x -17/18)^2 + (y - 17/18)^2 = r^2
%(y - 71/18)^2 = r^2 - (x - 17/18)^2
z = (x - 5);
real_z1 = z.*z;
z = real_z1;
y1 = 14 + sqrt((4 - z));
real_y1 = 14 + sqrt((4 - z));
y1 = real_y1;
y = 14 - sqrt((4 - z));
real_y2 = 14 - sqrt((4 - z));
y2 = real_y2;
plot (x, y3, x, y1, x, y2);
```

PROBLEM 289.

FIND THE EQUATION OF THE CIRCUMFERENCE CIRCUNSCRIT AT THE TRIANGLE DETERMINED BY THE STRAIGTH LINES $x + y - 2 = 0$, $9x + 5y - 2 = 0$ AND $y + 2x - 1 = 0$.

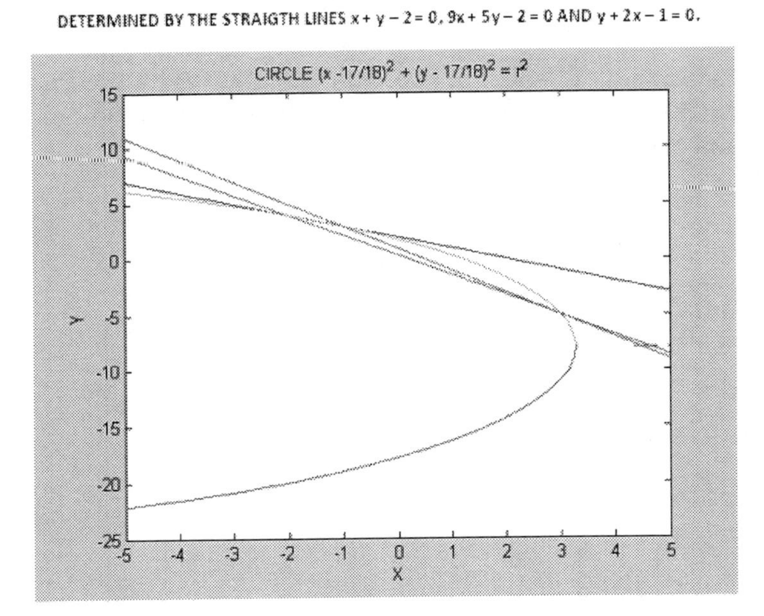


```
x = 3: .001: 7;
y3 = 2*x + 4;
x = 3: .001: 7;
%(x -17/18)^2 + (y - 17/18)^2 = r^2
%(y - 71/18)^2 = r^2 - (x - 17/18)^2
z = (x - 5);
real_z1 = z.*z;
z = real_z1;
y1 = 14 + sqrt((4 - z));
real_y1 = 14 + sqrt((4 - z));
y1 = real_y1;
y = 14 - sqrt((4 - z));
```

```
real_y2 = 14 - sqrt((4 - z));

y2 = real_y2;

plot (x, y3, x, y1, x, y2);
```

PROBLEM 290.

DETERMINE THE EQUATION OF THE CIRCUMFERENCE INSCRIT AT THE TRIANGLE DETERMINED BY THE STRAIGTH LINES $x + y = 1$, $y - x = 1$, $x - 2y = 1$.

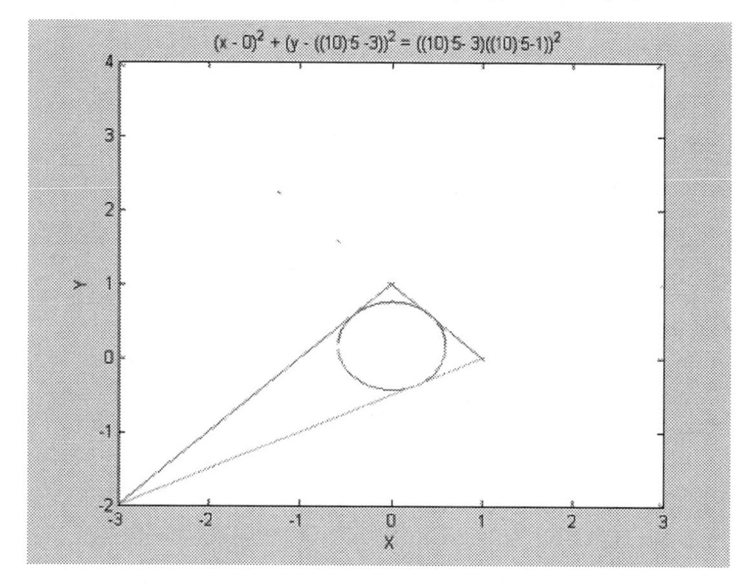


```
r = ((10)^.5 - 3)*((10)^.5 - 1);

r = r^.5;

x = r: -.001: - r;

%(x - 0)^2 + (y - ((10)^.5 -3))^2 = ((10)^.5- 3)((10)^.5-1));

z = (x);

real_z1 = z.*z;

z = real_z1;

y1 = (10)^.5 -3 + sqrt((r.*r - z));

real_y1 = (10)^.5 -3 + sqrt((r.*r - z));
```

```
y1 = real_y1;
y = (10)^.5 - 3 - sqrt((r.*r - z));
real_y2 = ((10)^.5- 3) - sqrt((r.*r - z));
y2 = real_y2;
%y = x + 1
x3 = -3: .001: 3;
y3 = 1 + x3;
%y = x/2 - 1/2;
x4 = -3: .001: 3;
y4 = (1/2)*x4 - 1/2;
x5 = -3: .001: 3;
y5 = -x5 + 1;
plot (x, y1, x, y2, x3, y3, x4, y4, x5, y5);
xlabel('X');
ylabel('Y');
title('(x - 0)^2 + (y - ((10)^.5 -3))^2 = ((10)^.5- 3)((10)^.5-1))^2 ');
```

PROBLEM 291. FIND THE ROOT LOCUS OF THE POINTS WHISH ARE OF THE SAME LONGITUDE THE TANGENTS AT THE CIRCUMFERENCE x^2 + y^2 = 4 AND x^2 + y^2 + 4y -2x = 4.

SUBSTRACTING ONE EQUATION FROM THE ANOTHER WE FIND THE ROOT LOCUS OF THE TANGENTS.

(x^2 + y^2 – 2x + 4y = 4) + (-x^2 - y^2 = -4) WE OBTAIN X = 2y

PROBLEM 292. SUBTRACTING THE EQUATION x^2 + y^2 -2x -2y = 0 FROM x^2 + y^2 + 2x – 6y +2 = 0 WE OBTAIN THE EQUATION OF ONE STRAIGTH LINE.

PROBLEM 293. DEMOSTRATE THAT THIS STRAIGTH LINE IS COMMON BOTH EQUATIONS REPRESENTED BY THE TWO FIRST EQUATIONS.

```
x^2 + y^2 – 2x - 2y = 0

- x^2 – y^2 - 2x + 6y - 2 =0;

-4x + 4y -2 = 0, x = y - ½

r = (3/16) + 12;

x = (3^.5)/4

x = x + r: -.001: -x - (r);

% 3/16 + 8(x - ((3)^.5)/4)^2 + 4(x - 4)^2

z = ((.5)^.5)*(x - 4);

real_z1 = z.*z;

z = real_z1;

y1 = (3^.5)/4 + sqrt((r + z));

real_y1 = (3^.5)/4 + sqrt((r + z));

y1 = real_y1;

y = (3^.5)/4 - sqrt((r + z));

real_y2 = (3^.5)/4 - sqrt((r + z));

y2 = real_y2;

plot (x, y1, x, y2);

xlabel('X');

ylabel('Y');

title('HIPERBOL % 3/16 + 8(x - ((3)^.5)/4)^2 + 4(x - 4)^2 ');
```

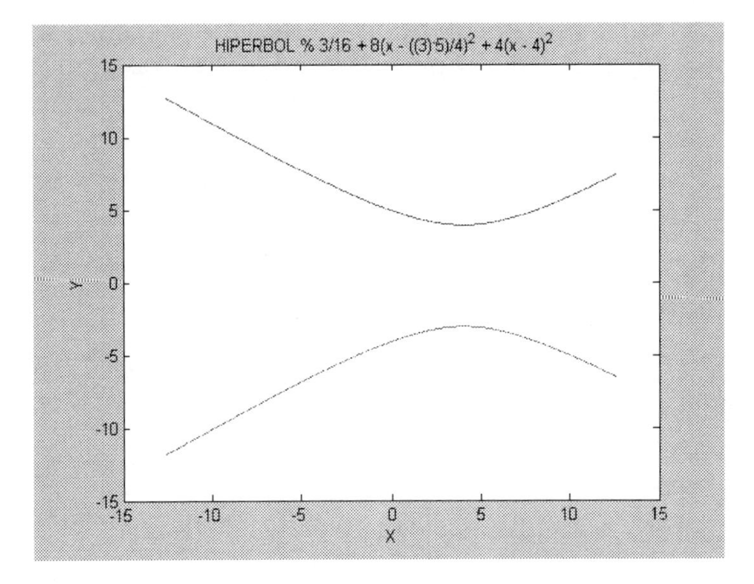
HIPERBOL % 3/16 + 8(x - ((3)·5)/4)^2 + 4(x - 4)^2
15
10
5
0
-5
-10
-15
Y
-15
-10
-5
0
5
10
15
X

ELLIPSE

BUILD THE GRAPHICS OF THE NEXT EQUATIONS, DEMOSTRATE THAT THE CURVES ARE ELLIPSE, FINDING HIS CENTERS AND SEMIAXLES.

294 X^2 + 2Y^2 = 6

(x – 0)^2 + 2(y – 0)^2 = 6 divide by 6 the equation

((x – 0)^^2)/6 + ((y – 0)^^2)/3 = 1 Center (0,0)

semi axle A = (6)^.5 semi axle B = (3)^.5

MATLAB SOLUTION

```
%x^2 + 2*y^2 = 6

z1 = (6)^.5;

z2 = (6)^.5;

x1 = - z1: .001: z2;

y = sqrt((- x1.^2 + 6)/2);

plot (x1 , y, x1, - y);

xlabel('X');

ylabel('Y');

title('ELIPSE ');
```

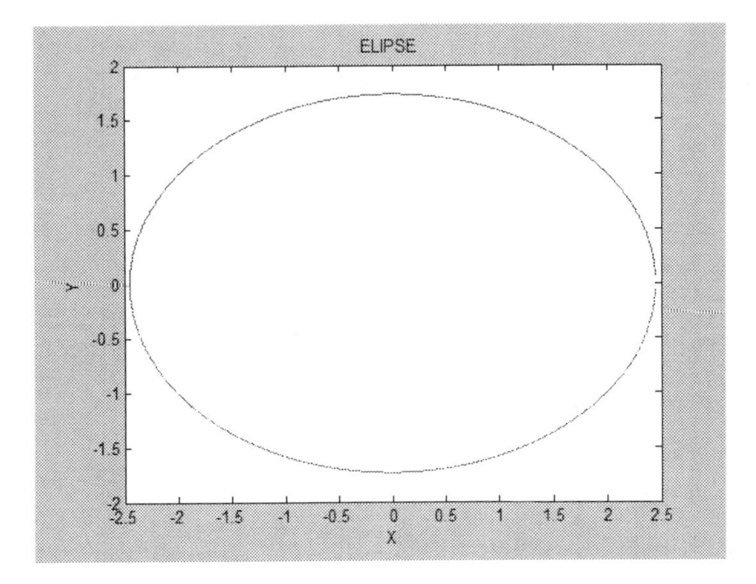


```
295 4x^2 + y^2 – 8x + 4y + 7 = 0
4(x^2 – 2x) + (y^2 + 4y) + 7 = 0
4(x – 1)^2 + (y + 2)^2 - 4 – 4 + 7 = 0
((x - 1)^2)/(1/4) + (y + 2)^2 = 1    C (1, - 2)    semi axle A = (1/4) semi axle B = 1
X = 1/ 4: .001: - 1/ 4;
Y + 2 = 1 - ((x - 1)^2)/(1/4)
Z = ((x - 1).^2)/(1/4)
Y = - 1 + z
plot (x1 , y, x1, - y);
xlabel('X');
ylabel('Y');
title('ELIPSE ');
%((x - 1)^2)/(1/4) + (y + 2)^ 2 = 1
x = -(6)^.5: .001: (6)^.5;
z = (x - 1).*(x - 1);
real_z1 = 1/4 - z;
```

```
z1 = real_z1;
y = sqrt(3*(4*z1));
y1 = sqrt((4*z1));
real_y1 = -2 + sqrt((4*z1));
y1 = real_y1;
y = - sqrt((4*z1));
real_y2 = -2 - sqrt((4*z1));
y2 = real_y2;
plot (x, y1, x, y2);
xlabel('X');
ylabel('Y');
title('ELIPSE ');
```

p 296 x^^2 + y^^2 = x `+ y

(x^^2 – x) + 2(y^^2 –1/ 2) = 0

(x – 1/ 2)^2 + 2(y – 1/ 4)^2 - 1/ 4 – 2(1/16) = 0

(1/2)(x – 1/ 2)^2 + (y – 1/ 4)^2 = 6/32 = 3/ 16 divide both members by 3/ 8

((x – 1/ 2)^2)/(6/16) + (y – 1/ 4)^2/(3/ 16) = 1

((x – 1/ 2)^2)/(6/ 16) + (y – 1/ 4)^2/(3/ 16) = 1

C (1/ 2, 1/ 4) semiaxle A = (1/ 4)(6)^.5 semi axle B = (1/ 4)(3)^.5

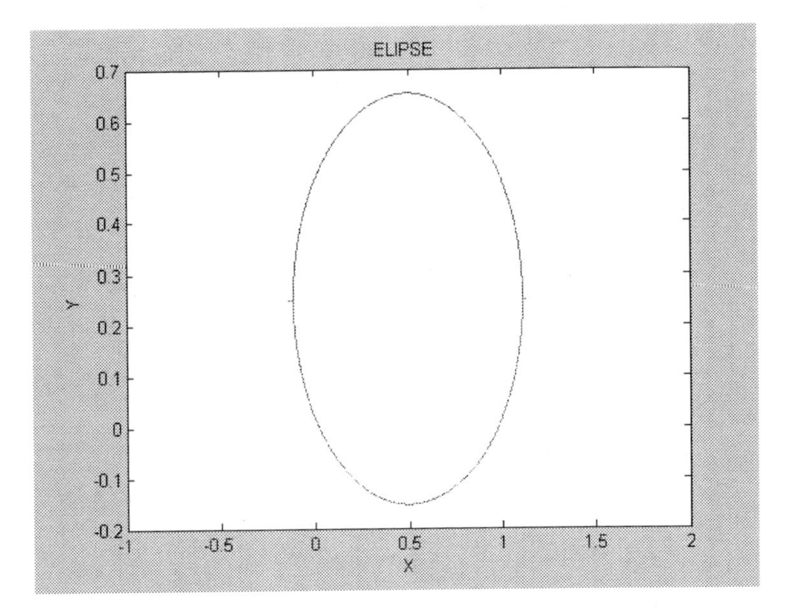


```
% ((x – 1/ 2)^2)/(6/ 8) + (y – 1/ 4)^2/(3/ 16) = 1
%((x - 1)^2)/(1/4) + (y + 2)^ 2 = 1
x = -.5*(6)^.5 +.5: .001: .5 +.5*(6)^.5;
z = (x - 1/ 2).*(x - 1/ 2);
real_z1 = 3/ 8 - z;
z1 = real_z1;
y = sqrt((3^.5/ 4)*z1);
y1 = sqrt((3^.5/ 4)*z1);
real_y1 = 1/4 + sqrt((3^.5/4)*z1);
y1 = real_y1;
y = - sqrt((3^.5)/(4)*z1);
real_y2 = 1/ 4 - sqrt((3^.5/4)*z1);
y2 = real_y2;
plot (x, y1, x, y2);
xlabel('X');
```

ylabel('Y');

title('ELIPSE ');

P 297 3x^2 + 2y^2 - 6x + 8y = 1

3 (x^2 – 2x) + 2 (y^2 + 4y) = 1

3 (x – 1)^^2 + 2(y + 2)^^2 - 3 - 8 = 1

3 (x – 1)^2 + 2(y + 2)^2 = 12 divide by 6

(1/ 2) (x – 1)^2 + (1/ 3)(y + 2)^2 = 2 divide by 2

(1/ 4)(x – 1)^2 + (1/ 6)(y + 2)^2 = 1 C(1, -2) semiaxle A = 2 semi axle B =(6)^.5

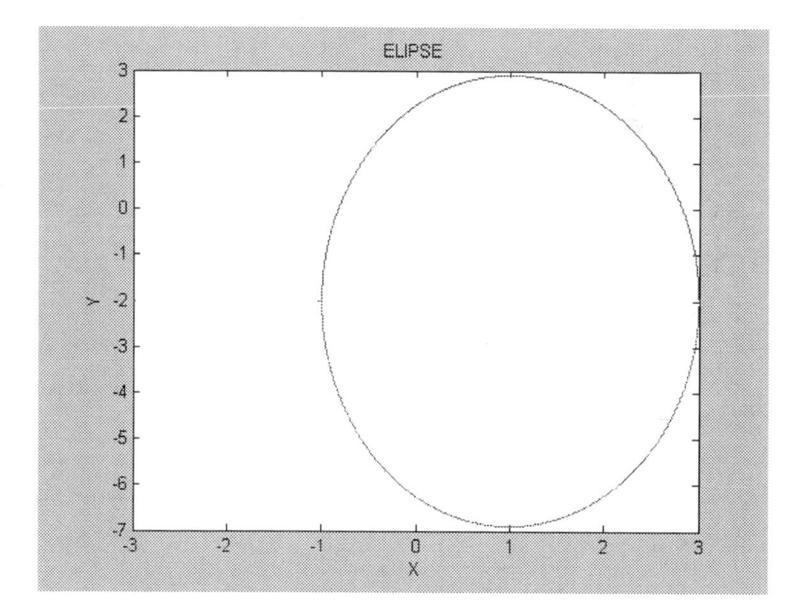

%(1/ 4)(x - 1)^2 + (1/ 6)(y + 2)^2 = 1

%C(1, -2) semiaxle A = 2 semi axle B =(6)^.5

x = -1 - 2: .001: 1 + 2;

z = (x - 1).*(x - 1);

real_z1 = 4 - z;

z1 = real_z1;

y = sqrt((6)*z1);

```
y1 = sqrt((6)*z1);
real_y1 = -2 + sqrt((6)*z1);
y1 = real_y1;
y = - sqrt((6)*z1);
real_y2 = -2 - sqrt((6)*z1);
y2 = real_y2;
plot (x, y1, x, y2);
xlabel('X');
ylabel('Y');
title('ELIPSE ');
```

P 298 x^^2 + 3y^^2 + 6x - 6y = 1

x^^2 + 6x + 3y^^2 - 6y = 1

(x + 3)^^2 + 3(y -1)^^2 - 9 - 3 = 1

(x + 3)^^2 + 3(y -1)^^2 = 13 divide by 3

(1/ 3)(x + 3)^2 + (y -1)^^2 = 13/3 divide by 13/ 3

(1/13)*(x`+ 3)^2 + ((y - 1)^2)/(13/ 3) = 1 C (-3, 1) semiaxle A = (13)^.5

semiaxle B = (13/3)^.5

p 299 5x^^2 + 2y^^2 - 10x + 4y +7 = 0

5(x^^2 - 2x) + 2(y^^2 + 2y) = -7

5(x – 1)^^2 + 2 (y + 1)^^2 - 5 -2 = -7

5(x – 1)^^2 + 2 (y + 1)^^2 = 0 Ellipse reduce at its Center.

PROBLEM 300. FIND THE EQUATION OF THE ELLIPSE HAVE CENTER C (1, -3) AND ITS AXLES PARALLEL OX AND OY ARE 2 AND 3 RESPECTIVELY.

((X – 1)^2)/4 + ((Y +3)^2)/9 = 1

PROBLEM 301. FIND THE EQUATION OF THE ELLIPSE HAVE HOW CENTER (-2, 4) AND IS TANGENT AT THE TWO COORDINATE AXLES.

```
x = -4: .001: 0;
z = (x + 2);
real_z1 = z.*z;
z = real_z1;
y2 = 4 + 2*sqrt((4 - z));
y1 = 4 - 2*sqrt((4 - z));
real_y1 = 4 - 2*sqrt((4 - z));
y1 = real_y1;
y2 = 4 + 2*sqrt((4 - z));
real_y2 = 4 + 2*sqrt((4 - z));
y2 = real_y2;
plot (x, y1, x, y2);
```

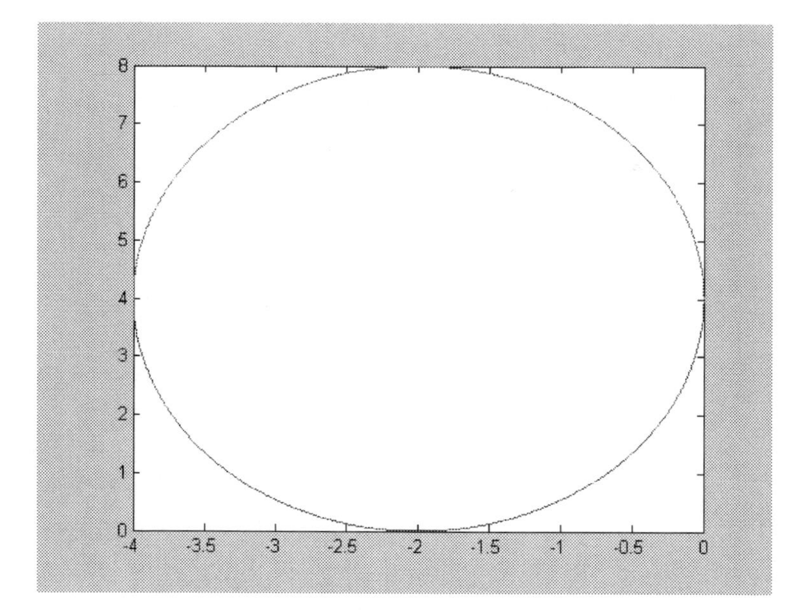

PROBLEM 302. FIND THE EQUATION OF THE ELLIPSE WHOSE AXLES ARE THE STRAIGTH LINES $x + y - 2 = 0$, $x - y + 2 = 0$ AND SEMIAXLES WITH LONGITUDE 1 AND 4 RESPECTIVELY.

NP = (X + y -2)/(2)^.5 MP = (x –y +2)/(2)^.5

NP^2/1 + MP^2/16 = 1

((x + y -2)^2)/2 + ((x-y +2)^2)/32 = 1

PARABOL

BUILD THE GRAPHICS OF THE NEXT EQUATIONS, DEMOSTRATE THAT ARE PARABOLS AND FIND IT'S AXLES AND VERTEXS.

PROBLEM 306. $Y^2 = 8x - 4$

$(y - 0)^2 = 8(x - 1/2)$ Vertex (1/ 2, 0) Axle $y = 0$

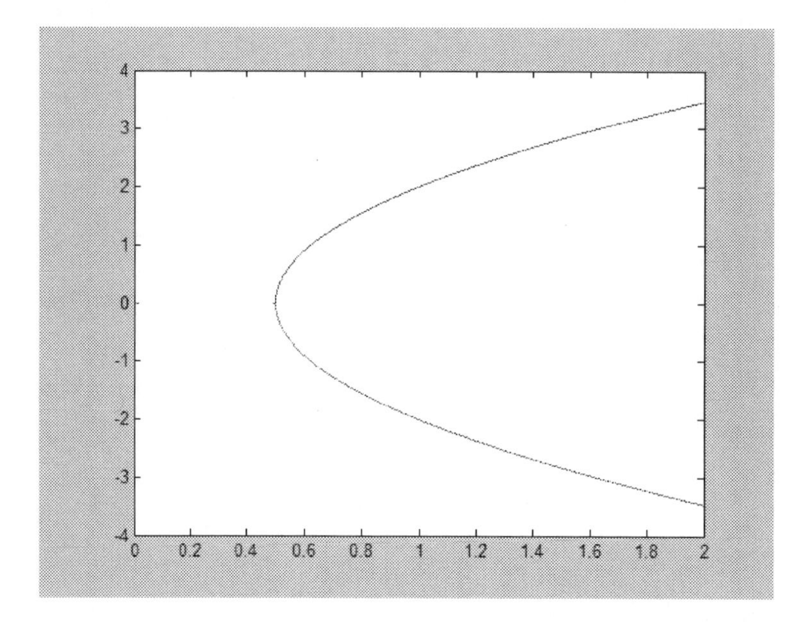

MATLAB PROGRAM

```
x = 0: .001: 2
y = 2*sqrt(2*x - 1);
y1 = - 2*sqrt(2*x - 1);
plot (x, y, x, y1)
```

PROBLEM 307. Y ^2 = - 2x + 1

(y – 0)^2 = -2(x – 1/ 2) Vertex (1/ 2, 0) axle Y = 0

MATLAB PROGRAM

```
x = -2: .001: .5;
y = sqrt (1 - 2*x);
y1 = - sqrt(1 - 2*x);
plot(x,y,x,y1);
```

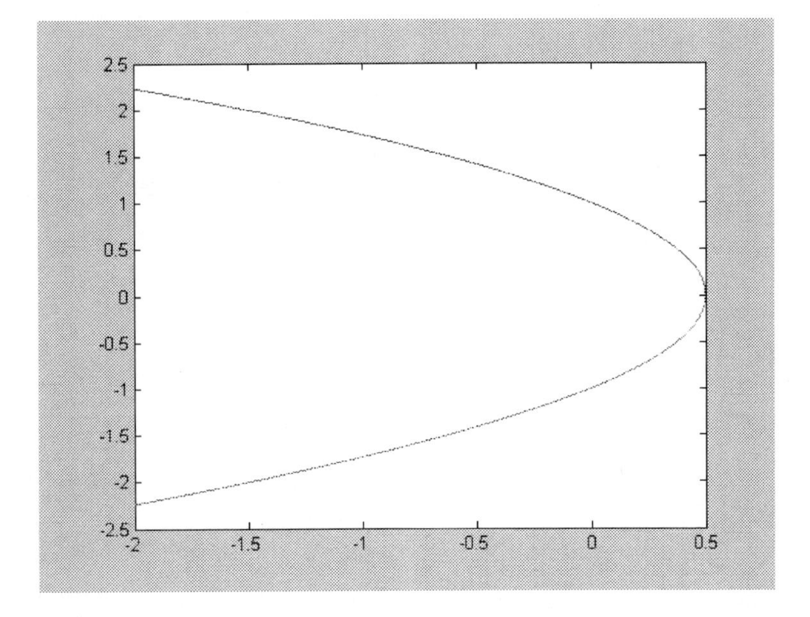

PROBLEM 308. Y = x^^2 – 2x + 3

y - 2= (x – 1)^^2 Vertex (1, 2) Axle x = 1

MATLAB PROGRAM

```
y = 2: .001: 5;
x = 1 + sqrt(y -2);
x1 = 1 - sqrt(y - 2);
plot(x, y, x1, y);
```

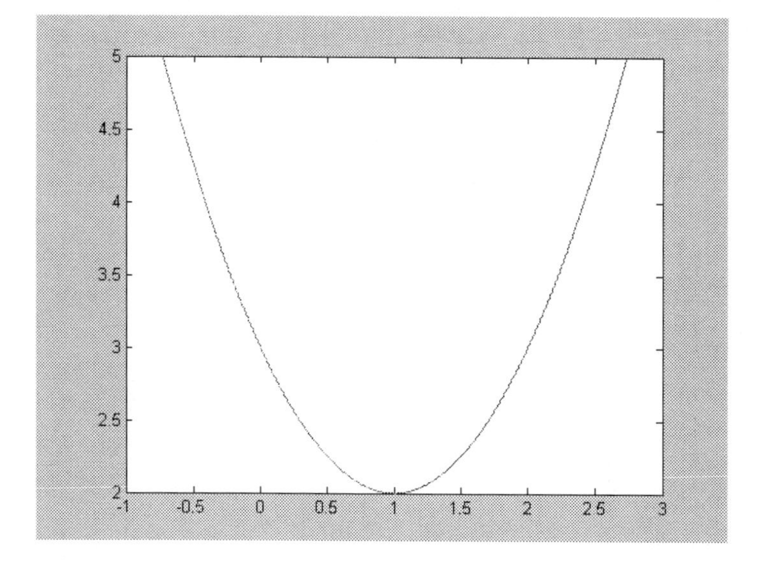

PROBLEM 309. $Y = (x - 1)(x + 2)$

$y = x^2 + x - 2$

$y = (x + 1/2)^2 - 9/4$

$(y + 9/4) = (x + 1/2)$ Vertex $(-1/2, -9/4)$ axle $x = -1/2$

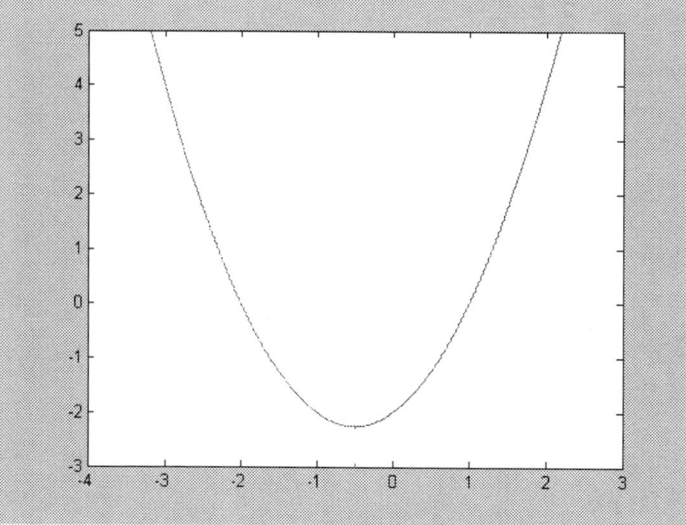

MATLAB PROGRAM

```
y = -3: .001: 5;
x = -.5 + sqrt(y + 9/4);
x1 = -.5 - sqrt(y + 9/4);
plot(x, y, x1, y);
```

PROBLEM 310. X = y^^2 - 3y

x = (y – 3/ 2)^2 - 9/ 4

x + 9/ 4 = (y – 3/ 2)^2 Vertex (-9/ 4, 3/ 2) axle y = 3/ 2

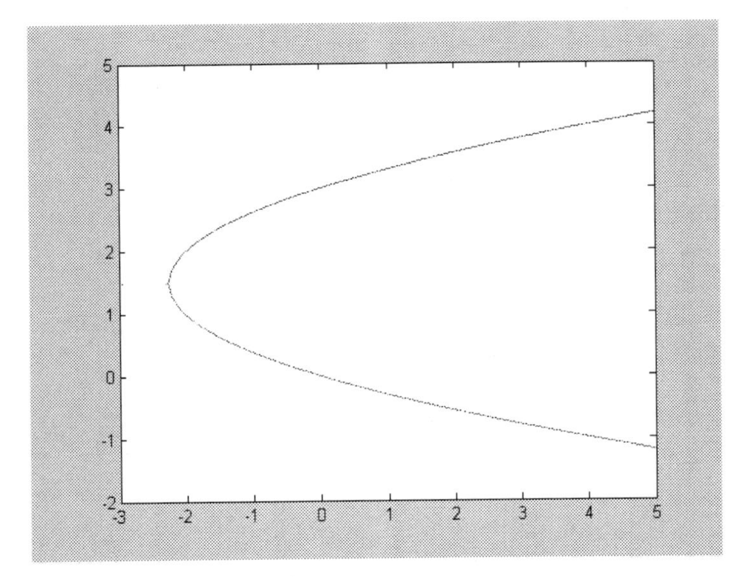

MATLAB PROGRAM.

```
x = -3.0: .001: 5;
y = 1.5 + sqrt(x + 9/4);
real_y = y;
y1 = 1.5 - sqrt(x + 9/4);
y = real_y;
plot(x, y, x, y1);
```

PROBLEM 312. FIND THE EQUATION OF THE PARABOL PASS BY (3, 4), IT HAS ITS HORIZONTAL AXLE AND ITS VERTEX IN THE COORDINATES ORIGIN.

(Y- 0)^2 = P (X -0)

16 = P(3) P = 16/ 3

(Y- 0)^^2 = (!6/3) (X – 0)

PROBLEM 313. FIND THE EQUATION OF THE PARABOL PASS BY (1, -3) HAVE IT'S VERTICAL AXLE AND VERTEX IN THE POINT (-2, 2).

```
y = 0: .001: 2;
x = -2 + sqrt((4/5 - (2*y)/5));
real_x = -2 + sqrt((4/5 - (2*y)/5));
x = real_x;
x1 = -2 - sqrt((4/5 - (2*y)/5));
real_x1 = -2 - sqrt((4/5 -(2*y)/5));
x1 = real_x1
plot (x, y, x1, y);
```

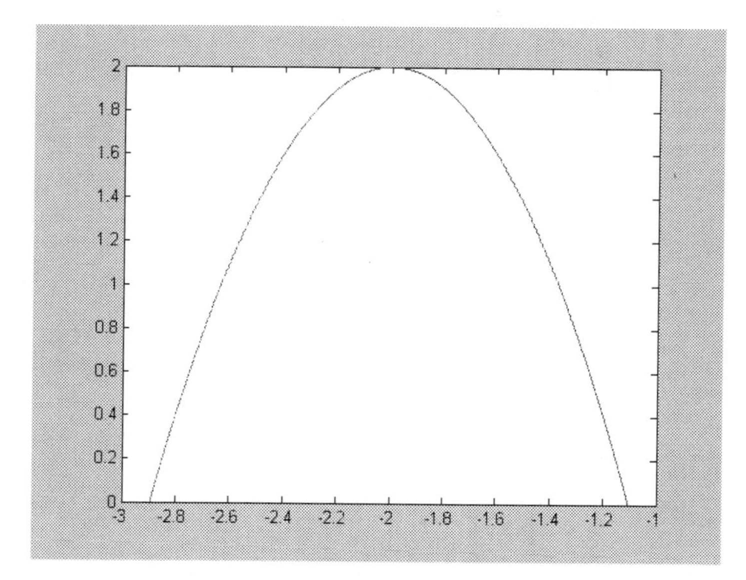

PROBLEM 315. THE CABLE OF ONE BRIDGE HAS BEEN SUSPENDED IN THE WAY OF ONE PARABOL WITH VERTICAL AXLE. THE HORIZONTAL AXLE HAS A WIDE OF 240 FEET, IT IS HELD BY ONE SYSTEM OF VERTICAL CABLES JOINED AT THE SUSPENSION, WHICH THE LONGEST MEASURE 80 FEET AND THE SHORTEST 30 FEET. FIND THE LONGITUDE OF THE VERTICAL CABLE OF THE HORIZONTAL AXLE AT ONE DISTANCE OF 40 FEET OF THE MIDLE POINT OF THE SAME.

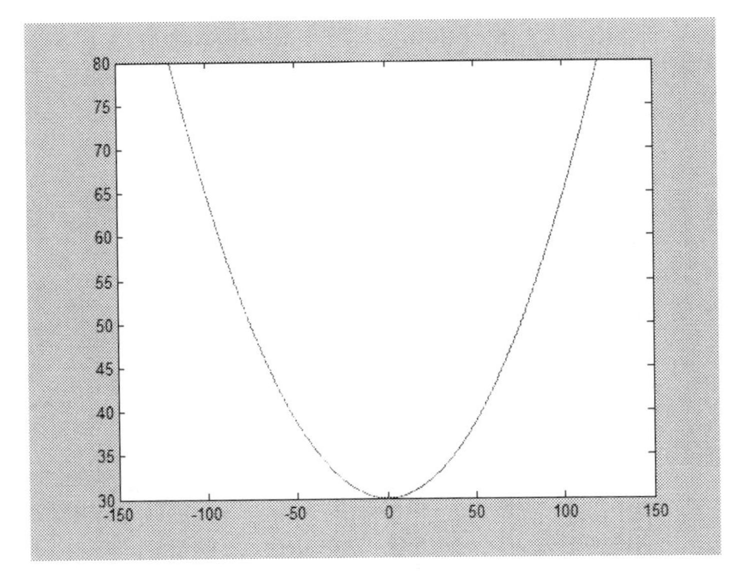

HIPERBOLE

DEMOSTRATE THAT THE NEXT EQUATIONS REPRESENT HIPERBOLES AND FIND IT'S CENTERS, AXLES AND ASIMPTOTHES.

PROBLEM 318.

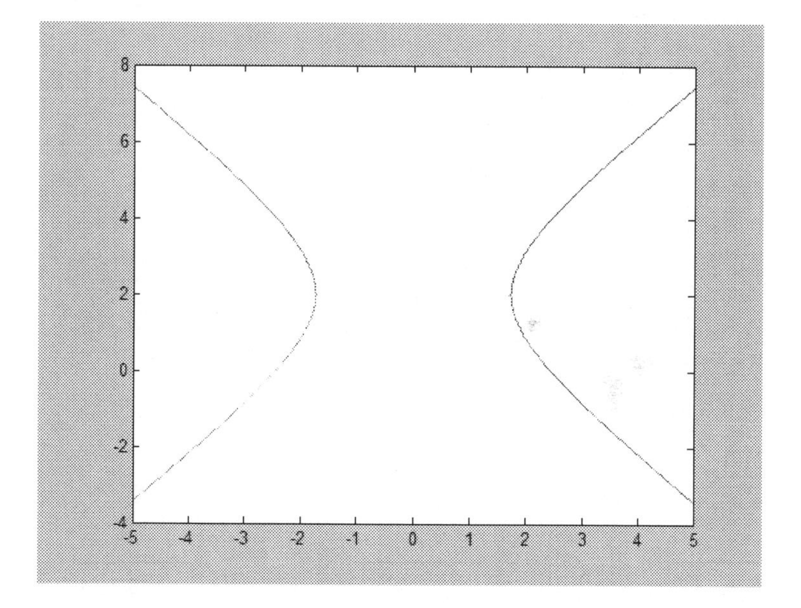


```
x = 1: .001: 5;

y = 2 + sqrt(4*(x.*x)/3 - 4);

real_y = 2 + sqrt(4*(x.*x)/3 - 4);

y = real_y;

y1 = 2 - sqrt(4*(x.*x)/3 - 4);

real_y1 = 2 - sqrt(4*(x.*x)/3 - 4);
```

```
y1 = real_y1;
plot(x, y, x, y1, -x, y, -x, y1);
```

PROBLEM 319.

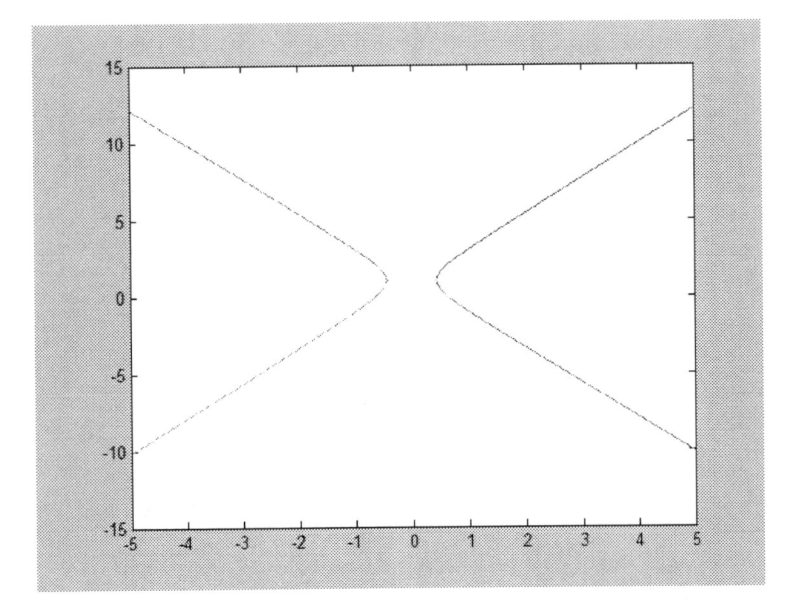


```
x = 0: .001: 5;
y = 1 + sqrt(5*(x.*x) - 1);
real_y = 1 + sqrt(5*(x.*x) - 1);
y = real_y;
y1 = 1 - sqrt(5*(x.*x) - 1);
real_y1 = 1 - sqrt(5*(x.*x) - 1);
y1 = real_y1;
plot(x, y, x, y1, -x, y, -x, y1);
x = 0: .001: 5;
y = 1 + sqrt(5*(x.*x) - 1);
real_y = 1 + sqrt(5*(x.*x) - 1);
```

```
y = real_y;
y1 = 1 - sqrt(5*(x.*x) - 1);
real_y1 = 1 - sqrt(5*(x.*x) - 1);
y1 = real_y1;
plot(x, y, x, y1, -x, y, -x, y1);
```

PROBLEM 321

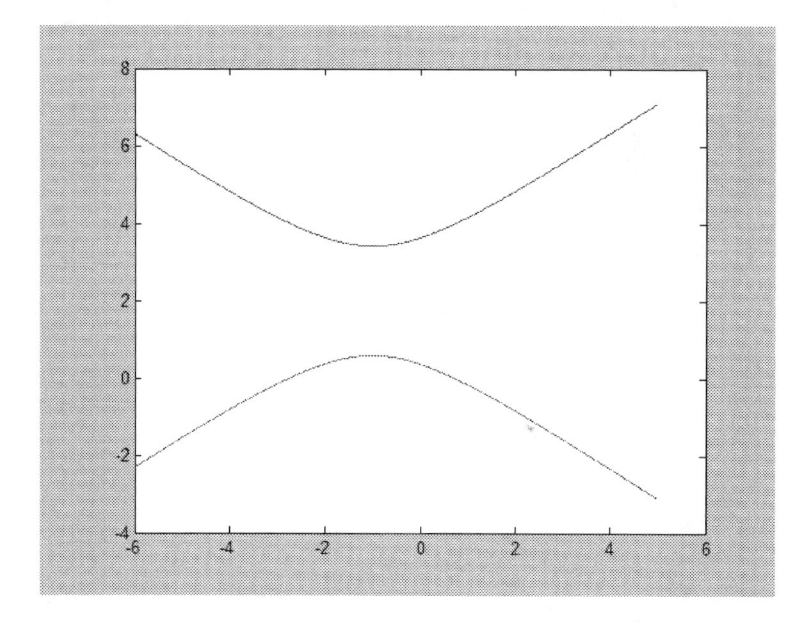


```
x = -6: .001: 5;
real_y = 2 + sqrt(2*(1 + (x + 1).*(x + 1)/3));
y = real_y;
real_y1 = 2 - sqrt(2*(1 + (x +1).*(x + 1)/3));
y1 = real_y1;
plot (x,y,x,y1);
```

ROOT LOCUS PROBLEMS

PROBLEM 349. ONE POINT MOVES SUCH WAY, THE SUM SQUARES OF THE DISTANCES AT THE FOUR SIDES OF ONE SQUARE, IS DOUBLE THE AREA OF THE SQUARE. FIND THE ROOT LOCUS OF THE POINT.

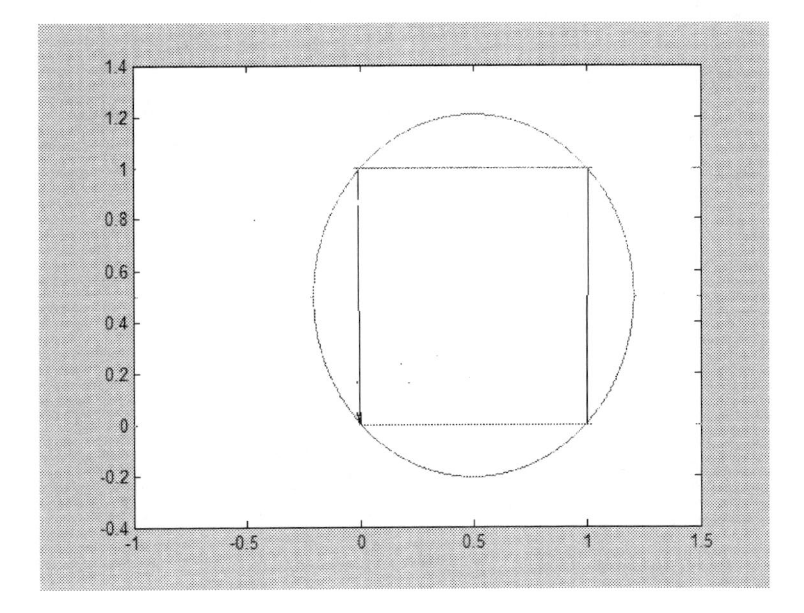


```
x = -1: .001: 1.5;

y = .5 - sqrt(.5 -(x - .5).*(x - .5));

real_y =.5 - sqrt(.5 -(x - .5).*(x - .5));

y = real_y;

y1 = .5 + sqrt(.5 -(x - .5).*(x - .5));

real_y1 = .5 + sqrt(.5 -(x - .5).*(x - .5));
```

```
y1 = real_y1;
x = -1: .001: 1.5;
y2 =1;
x = -1: .001: 1.5;
y3 = 0;
plot(x, y, x, y1, x, y2, x, y3);
```

GRAPHIC AND EMPIRIC EQUATIONS

GRAPHIC REPRESENTATION OF EQUATIONS.

PROBLEM 362.

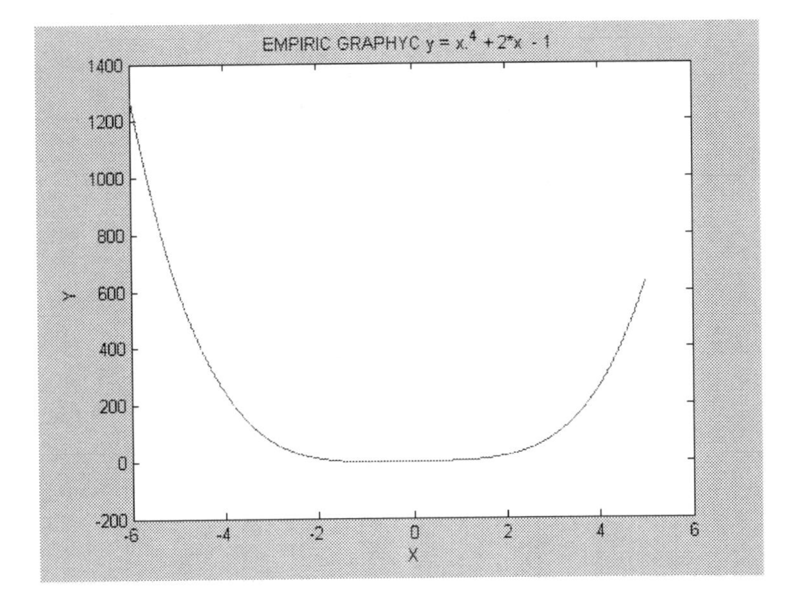


```
x = -6.0: .001: 5.0;
y = x.^4 + 2*x - 1
plot (x,y);
xlabel('X');
ylabel('Y');
title(' EMPIRIC GRAPHYC y = x.^4 + 2*x - 1 ');
```

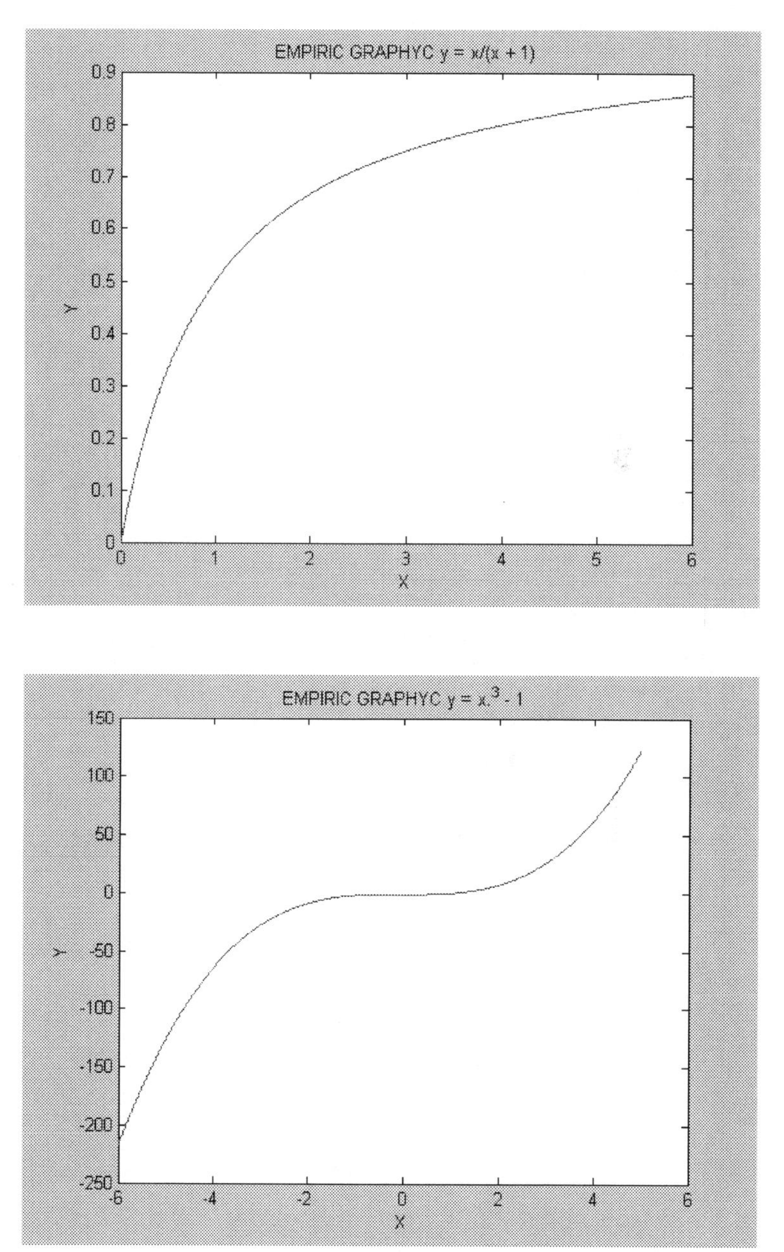
EMPIRIC GRAPHYC y = x/(x + 1)
0.9
0.8
0.7
0.6
0.5
0.4
0.3
0.2
0.1
0
Y
0 1 2 3 4 5 6
X
EMPIRIC GRAPHYC y = x.^3 - 1
150
100
50
0
-50
-100
-150
-200
-250
Y
-6 -4 -2 0 2 4 6
X

```
x = -6.0: .001: 5.0;
y = x.^3 - 1
plot (x,y);
xlabel('X');
ylabel('Y');
title(' EMPIRIC GRAPHYC y = x.^3 - 1 ');
```

POLAR EQUATIONS

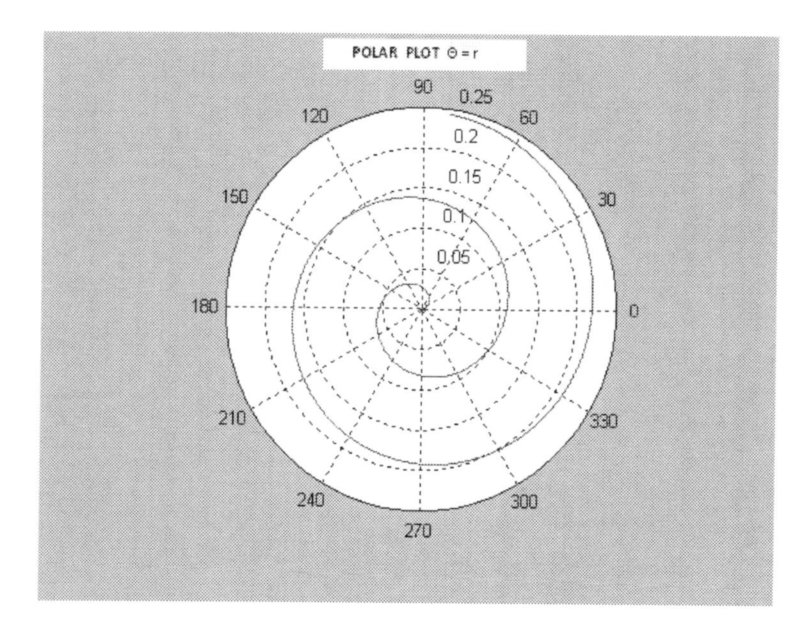


```
r = 0.: .001: 14.0;
THETA = (r)*pi/180;
polar(r, THETA)
title ('POLAR PLOT THETA = r');
```

PARAMETRIC EQUATIONS

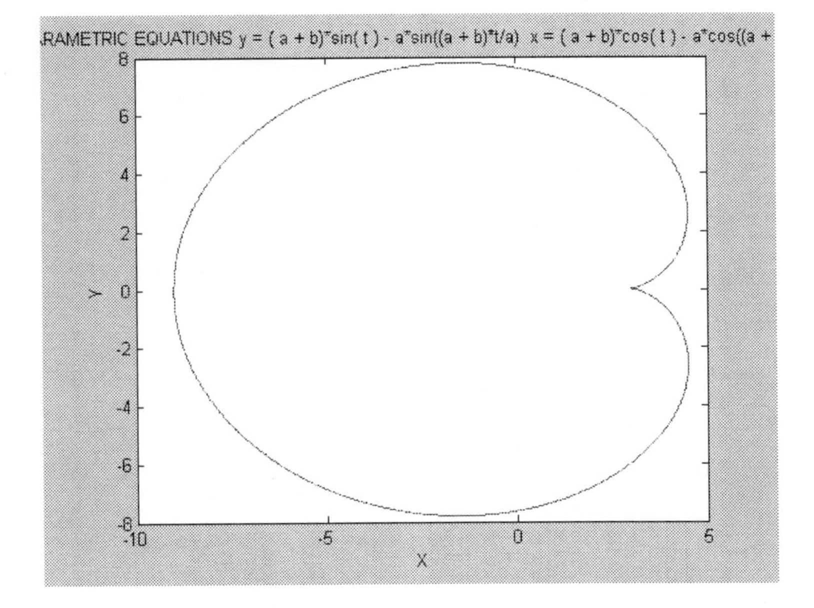


```
t = -4*pi: .001*pi: 4*pi;
a = 3;
b = 3;
y = (a + b)*sin(t) - a*sin((a + b)*t/a);
x = (a + b)*cos(t) - a*cos((a + b)*t/a);
plot (x,y);
xlabel('X');
ylabel('Y');
```

```
title(' PARAMETRIC EQUATIONS y = (a + b)*sin(t) - a*sin((a + b)*t/a) x = (a +
b)*cos(t) - a*cos((a + b)*t/a)');
```

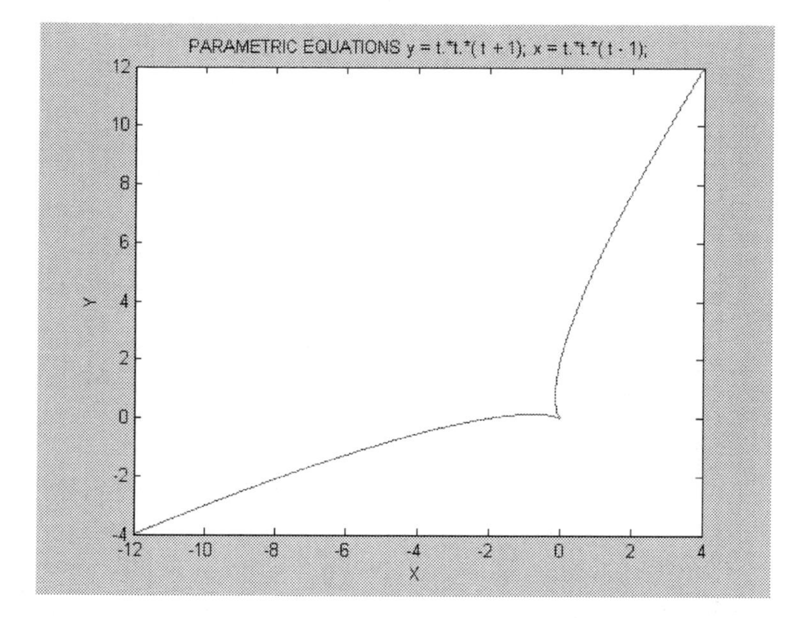


```
t = -2.0: .001: 2.0;
y = t.*t.*(t + 1);
x = t.*t.*(t - 1);
plot (x,y);
xlabel('X');
ylabel('Y');
title(' PARAMETRIC EQUATIONS y = t.*t.*(t + 1); x = t.*t.*(t - 1); ');
```

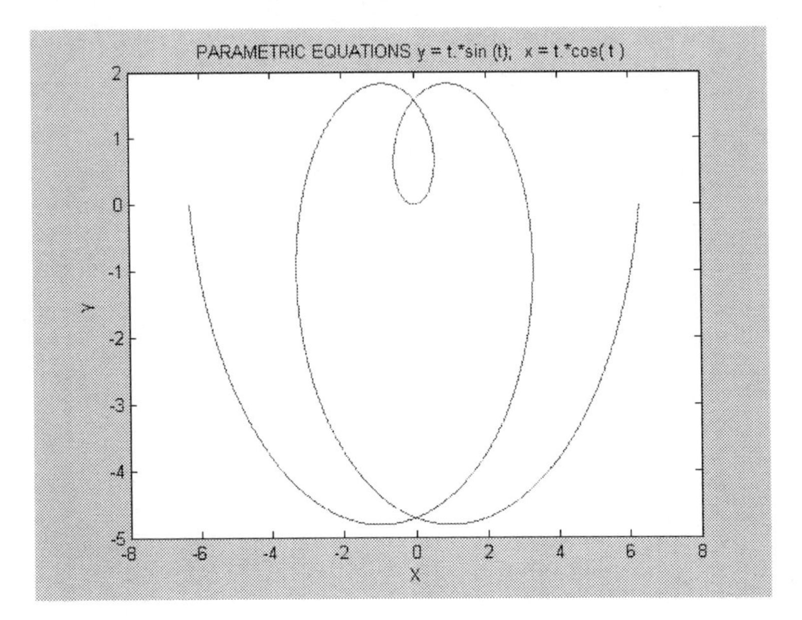


```
t = -2*pi: .001*pi: 2*pi;
y = t.*sin (t);
x = t.*cos(t);
plot (x,y);
xlabel('X');
ylabel('Y');
title(' PARAMETRIC EQUATIONS y = t.*sin (t); x = t.*cos(t) ');
```

BIBLIOGRAPHY

GEOMETRIA ANALITICA — H. B. PHILLIPS

TRIGONOMETRIA PLANA Y ESFERICA — WEBSTER WELLS S.B.

THE STUDENT EDITION OF MATLAB THE LANGUAGE OF THECNICAL COMPUTING